Bioassays with Arthropods, Third Edition

Bioassays with Arthropods, Third Edition

Jacqueline L. Robertson
Moneen M. Jones
Efren Olguin
Brad Alberts

CRC Press
Taylor & Francis Group
Boca Raton London New York

CRC Press is an imprint of the
Taylor & Francis Group, an **informa** business

CRC Press
Taylor & Francis Group
6000 Broken Sound Parkway NW, Suite 300
Boca Raton, FL 33487-2742

Printed on acid-free paper

International Standard Book Number-13: 978-1-4822-1708-7 (Hardback)

Library of Congress Cataloging-in-Publication Data

Names: Robertson, Jacqueline L.
Title: Bioassays with arthropods / Jacqueline L. Robertson [and three others].
Description: Third edition. | Boca Raton : CRC Press, 2017. | Includes bibliographical references.
Identifiers: LCCN 2016053157| ISBN 9781482217087 (hardback : alk. paper) | ISBN 9781482217094 (ebook pdf)
Subjects: LCSH: Pesticides--Environmental aspects--Measurement. | Arthropoda--Effect of pesticides on. | Biological assay.
Classification: LCC QH545.P4 R478 2017 | DDC 363.17/92--dc23
LC record available at https://lccn.loc.gov/2016053157

Visit the Taylor & Francis Web site at
http://www.taylorandfrancis.com

and the CRC Press Web site at
http://www.crcpress.com

Dedication for the third edition

This edition is dedicated to Jacqueline L. Robertson, my mentor and friend, whose support through my doctorate and early career helped me understand what a colleague should be.

Moneen M. Jones

From the second edition

To David J. Finny of the University of Edinburgh, whose work is the basis of all that we have done.

From the first edition

To Lucille Boelter (1921–1988), my technician and friend: she did so much of my work that I had time to think, and to Johnne Nicholas (1915–2000), whose moral support meant so much to me.

Jacqueline L. Robertson

Contents

Preface

The third edition of *Bioassays with Arthropods* is a tribute to my friend and colleague, Jacqueline L. Robertson.

Jackie and I established a mentor/shadow relationship while I was a doctoral student attending University of Illinois, Urbana-Champaign. My research emphasis was screening Oriental fruit moth for resistance to conventional and reduced risk pesticides. I picked up a copy of *Bioassays with Arthropods* (2nd Edition) to learn all I could about bioassays and the PoloPlus software. Many times, I called Jackie to discuss detailed explanations for my research results, and before I knew it, she was a member of my dissertation committee. Each chapter of my dissertation involved conducting bioassays. I learned to use PoloPlus for estimating dose–response curves for conventional and reduced-risk insecticides and insect growth regulators. Now, when I say I learned how to do bioassays, I must say that I was conducting bioassays in excess of 1200 individual larvae a week for 3 years. I learned to be meticulous when analyzing the results. Jackie's newly developed software, PoloMix, was used to analyze my dissertation data to determine the occurrence of synergism or antagonism in tank mixes of chemicals. It was the first time we had published data using this software. The doctorate research culminated with four publications, with Jackie being second author on each. Jackie and I continued a professional and personal friendship following my graduation. In fact, she added me (2010) to her list of employees for LeOra Software company as an entomologist consultant.

Similar to Jackie, I had no intention of becoming an entomologist. In 2006, I completed dual masters, one in biology and the other in geography and environmental studies. I could have easily become a Geographic Information System specialist researching temporal and spatial analyses of data if it had not been for an opening for a tree crop PhD assistantship at the University of Illinois, Urbana-Champaign. My acceptance into that doctoral program introduced me to integrated pest management and toxicology. With thousands of bioassays (no exaggeration) under my belt, I am happy I made the decision to study entomology. I am also thrilled that I made Jackie's acquaintance through her book.

Because of all the questions I had while reading Jackie's book as a graduate student (and there were plenty, I can assure you), Jackie and I agreed that they could be answered in the third edition of the book. I literally saved hundreds of e-mails between the two of us while I was a student. In addition, we could use my dissertation data for examples. Jackie contracted with the publisher in the middle of 2013 for a revised edition of the book to be published early 2015. However, when Jackie abruptly passed away in July of 2014, it was a crushing shock to her friends and colleagues. A year passed before I could return to writing the chapters. I apologize to all those that were patiently waiting for this book.

This third edition was written to clarify paragraphs and examples written in the second edition. A little background on how the programming for probit analysis evolved is offered here. The first software, written by Robert Russell, was POLO,[1] a probit or logit analysis program that ran on a Univac 1108 series mainframe computer that was like something out of *Desk Set.** Every detail of the probit program, down to the last digit, matched the output of David Finney's program BLISS.[2] However, automatic tests of the hypotheses of parallelism and equality were included, and the program itself reflected the latest developments in both statistics and computer sciences.[3] POLO led to POLO-PC,[4] which was specifically designed for personal computers. Next came the Microsoft Windows version, PoloPlus.[5] In PoloPlus, portions of the original output were omitted because there were just too many details to decipher that were important and that were extraneous details.

PoloPlus dealt with single variable bioassays, but sometimes, an entomologist wishes to sample multivariables, such as body weight with dose. The authors, Gene Savin (statistician), Robert Russell, Haiganoush Preisler (statistician), and Jackie Robertson, next tackled the question of the significance of body weight as a variable in bioassays. They released PoloEncore,[6] a vastly improved Windows version of this multivariate probit/logit program. As useful as multivariate probit analysis is, few entomologists have ever examined their assumptions about body weight or other explanatory variables besides dose. Although PoloEncore had good intentions to handle multivariate factors, it was very confusing how to enter the data, and it was laborious to input the data. The third-generation program, PoloMulti, reduces the complexities of the data entry with the program but maintains the robust data analyses.

The third edition also introduces the next generation of Polo software (i.e., PoloSuite) that was written for distribution by LeOra Software LLC. I had no access to the original code, nor did Jackie. Robert Russell, the original programmer of the Polo programs, first in Fortran then C, was unavailable. The code was forever lost. Through a mutual contact, I found Efren Olquin, a computer engineer of superb skill and intellect. He also likes a challenge. Did I mention that the original code was lost? We compared old data and new data with each software program module until the results matched perfectly. We also did some minor tweaking for ease of data entry. The book also had code that was written in GLIM, an old statistical programming language that is no longer available. A new statistician, Brad Alberts, was brought on board to rewrite the GLIM code to R and to update paragraphs in the book that could be written more succinctly. Our collaboration means Jackie's legacy of this book and software will not disappear for a very long time.

Jackie had a wicked sense of humor, as her first two editions of *Bioassay with Arthropods* demonstrated by the inclusion of characters in silly situations. I have attempted to continue the tradition by adding zany characters to the chapters. I also tried to use field and laboratory situations that have occurred in my research. I made obvious changes to each chapter by adding more details and explanations for why a method is chosen or done. Not every entomologist has years of statistics experience, so I tried to tone down the verbiage.

* In this 1958 MGM movie starring Katherine Hepburn and Spencer Tracy, a huge computer named EMARAK (designed by Spencer Tracy) computes (among other things) the total acreage defoliated by spruce budworm more quickly than do the research librarians (led by Katherine Hepburn). Several short circuits later, the computer blows up. Our UNIVAC 1100 was considerably more advanced, but about the same size.

The book is kept concise, with only short definitions of key concepts provided. Even though Jackie was an invaluable resource in answering complex bioassay questions, I still needed to consult textbooks and journal articles for details in toxicology and bioassays. I can assure you that reading *Bioassays with Arthropods* is a much easier read than David J. Finney when it comes to understanding probit analyses. For my dissertation research, I also enrolled in two upper division statistics courses through the advice of my graduate advisor, Rick Weinzierl. Suggested statistics books include *An Introduction to Statistical Methods and Data Analysis* by Ott and Longnecker[7] and *Design of Experiments: Statistical Principles of Research Design and Analysis* by Robert Kuehl[8] and for a crash course in toxicology, *Chemical Pesticides, Mode of Action and Toxicology* by Jørgen Stenersen.[9] While these books are not a substitute for a solid background in integrated pest management, statistics, or toxicology, they will supplement a reader's understanding of the material.

To this day, I can still hear Jackie's advice, comments, and critiques in my head whenever I begin writing. Whether it is a short extension article or lengthy entomology manuscript, Jackie, a former Entomological Society of America editor, had a way of burning corrections into your head so you never made the same grammar mistake twice. I am forever in her debt for those little tidbits of advice.

I hope you enjoy reading this book as much as I enjoyed writing it. I hope it will become an important reference guide for your research.

To order the PoloSuite computer software described in *Bioassays with Arthropods, Third Edition*, use the order form found at www.leora-software.com or contact the LeOra Software Company at leorasoftware@gmail.com.

Moneen M. Jones

REFERENCES

1. Robertson, J. L., Russell, R. M., and Savin, N. E., POLO: A user's guide to Probit or Logit analysis, *USDA Forest Service Gen. Tech. Rep.*, PSW-38, 1980.
2. Finney, D. J., *BLISS*, Department of Statistics, University of Edinburgh, Edinburgh, Scotland, 1971.
3. Russell, R. M. and Robertson, J. L., Programming probit analysis, *Bull. Entomol. Soc. Am.* 25, 191, 1979.
4. LeOra Software, POLO-PC, A user's guide to probit or logit analysis, 1997.
5. LeOra Software, PoloPlus, POLO for Windows, LeOra Software, 1007 B St., Petaluma, CA 94952.
6. LeOra Software, PoloEncore, LeOra Software, 1007 B St., Petaluma, CA 94952.
7. Ott, R. L. and Longnecker, M. T., *An Introduction to Statistical Methods and Data Analysis*, 6th ed., Cengage Learning, Boston, 2010.
8. Kuehl, R. O., *Design of Experiments: Statistical Principles of Research Design and Analysis*, 2nd ed., Cengage Learning, Boston, 1999.
9. Stenersen, J., *Chemical Pesticides, Mode of Action and Toxicology*, CRC Press, Boca Raton, FL, 2004.

Authors

Jacqueline (Jackie) L. Robertson received both her BA (1969, Zoology) and PhD (1973, Entomology) from the University of California–Berkeley. She joined the US Forest Service as a temporary employee in 1966, became a research scientist at the Pacific Southwest Research Station in 1972, and left as a senior scientist in 1996. She is responsible for biological interpretation of statistical results for LeOra Software.

Dr. Robertson was the author of more than 150 technical publications. She was a member of the Entomological Society of America (for which she served as an editor of the *Journal of Economic Entomology* from 1981 through 1996) and the Entomological Society of Canada. Her research interests concerned forensic veterinary entomology and statistical models of insect population responses to toxins in their environment.

Moneen M. Jones received her BS (2002, biology) from Elmhurst College, MS and MA (2006, biology and geography and environmental science, respectively) from Northeastern Illinois University, and PhD (2010) from the University of Illinois, Urbana-Champaign. She is a member of the Entomological Society of America and is an assistant professor research of entomology at the University of Missouri. Her research concerns integrated resistance management of row crop pests. She is also responsible for biological interpretation of statistical results for LeOra Software, LLC (2016).

Efren Olguin received his BS (2003) from the Monterrey Institute of Technology in Mexico City, Mexico. He received an MS (2010) in industrial engineering and operations research from the Pennsylvania State University. He has worked in several information technology projects involving mathematical modeling and data analysis. Mr. Olguin retooled the PoloSuite software using the original equations, so that several of the modules were more user friendly.

Brad Alberts received his BA (2014) from the Department of Mathematics and Statistics from Washburn University. He received a Master's in Statistics (2016) from the University of Missouri. At the University of Missouri, he was a statistical research assistant and consultant for the Social Science Statistics Center.

CHAPTER **1**

Introduction

Mā´ven, mā´vin, n. [Yiddish, from late Hebrew *meven*]. An expert or connoisseur, often a self-proclaimed one.

Webster's Unabridged Dictionary

A *bioassay* is any experiment in which a living organism is used as a test subject. When a stimulus is applied, the organism responds. A bioassay provides a means to quantify the response or responses. In a general sense, a *pesticide* is any natural or synthetic substance or organism that harms an undesirable organism—a pest. *Pest* has no scientific definition. Instead, this designation is strictly a product of human activities and needs and is totally anthropocentric. Most pests are members of the invertebrate Phylum Arthropoda; within this phylum, most arthropod pests are members of the classes Insecta or Arachnida. Some pests interfere with production of food, fuel, or fiber by humans.

Nuisance pests, with which many of us are familiar, include cockroaches, ants, and gnats. Others, such as wasps, yellowjackets, mosquitoes, and some species of spiders, can cause severe allergic reactions in humans and other mammals.[1] The most subjectively defined group of pests includes those that are ugly or those that provoke a fear response. The category of insects is vague, but large quantities of pesticides are purchased each year[2] to control urban pests that breach our territory boundaries.

A single pest is never the problem. If it was, I would write about my research experiences with bedbugs, and how a nymph (I named him Henry) survived 6 weeks without food (i.e., a blood meal) in a Ziploc baggie after he hitchhiked on the bottom of my shoe to my home, or the best way to reduce swelling and itching from said bedbug bite. Instead, we study how a pest *population* (an interbreeding group of individuals of the same species[3]) is responsible for damage, disease transmission, or annoyance. *Pesticide bioassays* are experiments done with a pesticide to estimate the probability that a pest population will respond in the desired manner (e.g., die, become sterile, or, at the very least, suffer horribly) and so be made innocuous.

Principles of valid bioassay apply not just to pests but to beneficial organisms as well. The same methods used to evaluate effectiveness on target (pest) species also can be used to estimate safety to nontarget species, such as parasitoids or predators.

Beyond tests with arthropods, the statistical methods and principles described in this book are also applicable to any tests in which an agent (drug, pesticide, herbicide, radiation) or treatment (heat, cold) is tested on any living thing.

To illustrate some of the problems involved in a pesticide bioassay, consider a novice in the field, Dr. Paula Maven. She wants to test an insect growth regulator (IGR) as a possible chemical to reduce populations of some very large and dangerous Lepidoptera (*Patronius giganticus* Ubetterduck). The resulting new species, first discovered in Marin County, California, was given the common name *Sicilian Godfather* for its love of all Italian vegetables. After its release into the ecosystem, the new species spread through California's Central Valley and swiftly chewed its way into the southern United States. Clearly, the Godfather must be controlled. At stake is an important part of the national economy (e.g., salsa and pizza) that depends on tomato, bell pepper, olive, onion, and garlic production. Her assignment is to establish an integrated pest management program for tomatoes.

The crown jewel of the University of Schaeferville is its International Research Center, a renowned institution that recruits graduate students for their skills in competitive insect pinning, quick taxonomy, and precision sweep netting.

The purpose of Dr. Maven's first experiment is quite deceptively simple. She wants to find a dose of the IGR that will control the population or at least kill about 90% of the larvae from hell. The first problem is practical: How is she to keep the ginormous larvae alive in the laboratory long enough to find her answer? After four hours of collecting infested vegetable stems that fill eight large trash bags, she has a grand total of 55 mature caterpillars, and a nice sunburn. The stems are placed into screened containers, where Dr. Maven discovers that the caterpillars eat tomato plants *and* each other. That leaves 35 test subjects, each of which must be kept in a separate 25-cm-diameter Petri dish. Dr. Maven begins the experiment by diluting the IGR to the concentration that the manufacturer recommends for mosquito control (i.e., at least 90% mortality of the larvae).

A second major problem arises: Should the IGR be applied directly to the caterpillar, to the tomato leaves, or to the filter paper lining the Petri dish? Or, should she wait until the larvae mature to adults, have them mate, and test the IGR on eggs? Dr. Maven chooses direct application to the larvae. After all, the IGR contacts mosquitoes and all other targets listed on the label. If it is effective by contacting the other pest species, she reasons that it should be effective on the Godfather in the same way.

Next, how many caterpillars should she test? Dr. Maven decides to use 25 of the 35 survivors (the other 10 will be used to start a laboratory colony, just in case she decides to test eggs at a later date). Now, how should the chemical be applied? She chooses microapplication (a process by which a small, measured drop of the pesticide is put on each insect's body) because published literature suggests that many other researchers use that method. Should the IGR be applied in proportion to the weight of each larva? Certainly, that is how physicians prescribe medication, so it seems reasonable to apply the chemical to each caterpillar in the same way.

With only 25 caterpillars, Dr. Maven next wonders how many dilutions to make from the stock solution. Five points are necessary to estimate a good regression curve, so five would be the minimum doses needed. Given her intention to construct a dose–response relationship for the IGR, Dr. Maven needs to dilute the stock solution five times, in 10-fold increments. Use of water for the solution seems reasonable because mosquitoes are exposed to the chemical in that medium. However, using acetone will permit the chemical to more easily penetrate the insect cuticle[4] for topical application. She selects a commercially formulated product for combining with acetone.

As each caterpillar is weighed, Dr. Maven assigns it to a concentration so that five will be treated with the first concentration, the second five with the second concentration, and so on. To not contaminate weaker concentrations with stronger ones, she will begin with the lowest concentration and progress to the highest using the micropipetter. Dr. Maven next places a carefully measured drop (1 μg) of the IGR solution on each caterpillar's back.

She waits for something to happen. Will the response of each caterpillar be an all-or-nothing event? The literature suggests that it may not be. The caterpillar might curl up and die, it might molt to the pupal stage and be malformed, or it might grow. Nevertheless, Dr. Maven selects death as the criterion for effectiveness because that would be the intent of using the IGR to protect tomato plants. To make the data fit the dead versus alive classification, she reduces the effects that she observes into one or the other category. In this case, moribund individuals will be combined with dead individuals to keep the responses into just two categories.

The insects are all treated, and now she needs to decide how long to wait for a response. At two days, there could be no differences in treatments, and at 10 days, all larvae could be dead or had molted to adults. Not wishing to miss any significant event, a middle date is chosen. Seven days later, all of the caterpillars are "dead." What has Dr. Maven accomplished? Unfortunately, nothing. Can a dose–response relationship be estimated from her data? Certainly not. Among the obvious deficiencies in Dr. Maven's experiment are lack of replication, use of a stock concentration appropriate for species in a totally unrelated insect family (i.e., mosquito), lack of a control (treated with acetone only), and use of an untested assumption that the response will be in proportion to body weight. The main problem, however, is that a dose–response relationship could not have been established with 25 test subjects even if she had done everything else perfectly. Clearly, Paula Maven needs help!

Dr. Maven has attempted to do a bioassay as a way to quantify the response or responses of the organism. A *quantal response* is one that varies in relation to a measurable characteristic of the stimulus. *Quantal response bioassays* are done to estimate the relationship between the response (or responses) and the quantity or intensity of a stimulus. These bioassays can differ in terms of the *response variables* and the *explanatory variables* (independent variables) involved. Explanatory variables are the measurable characteristics of a stimulus or stimuli that cause a response or responses to vary. Response variables (the dependent variables) are the random outcomes of the experiment; they vary in relation to the stimulus (i.e., in relation to the independent variables). *Statistical models* are used to estimate the mathematical relationships between and among variables.

Quantal response bioassays are often done with pest species. When reduction of numbers of pests in a population is the goal, bioassays are done to estimate the probability that the pest population will respond in the desired manner—they will, for example, die, become sterile, or, at the very least, suffer horribly—and so be made innocuous.

Although most quantal response bioassays have been done to estimate the activity of pesticides (natural or synthetic substances that harm a pest), principles of valid bioassay also apply to any physical factor that can be manipulated to affect numbers of pests. Radiation (e.g., Lester and Barrington[5]), microclimate,[4] and temperature[6] have been used in some situations to minimize the use of or completely replace synthetic organic chemicals. Many procedures for organic gardening are also based on bioassays; these include estimating the amounts of active ingredients necessary for baits, attractants, and repellents (e.g., see Roberts[7]). Beneficial microorganisms such as bacteria (*Bacillus thuringiensis*),[8] insect viruses (nuclear polyhedrosis virus[9] and granulosis), nematodes,[10] and protozoa (*Nosema* spp.; e.g., Burges[11]) also used in organic gardening are tested in dose–response bioassays before they are developed commercially.

Van den Bosch's[12] principles of safety and selectivity in the environment mean that all organisms, not just pests, must be considered when any control method is used. Quantal response bioassays can and should be used to estimate risk to nontarget species, such as parasites, predators, and other organisms in the environment. Beyond the arthropods and in the general framework of environmental health and safety, the methods and principles described in this book are applicable to any tests where an agent or treatment is tested on any living thing.

Our primary emphasis is the design of valid quantal response bioassays and use of appropriate statistical models to analyze resultant data. The information and suggestions presented here should help even a novice like Paula Maven to avoid mistakes that can invalidate an experiment. Without good experimental design, data from a bioassay will be meaningless at worst and suspect at best. In any quantal response bioassay, experiments must be carefully designed so that maximum biological information can be extracted from statistically valid data. Design includes considerations such as identification of response variables and explanatory variables, numbers of replications, where to place doses in a simple dose–response bioassay to achieve greatest precision, and the number of subjects that must be tested. These topics were treated previously in statistical texts by Finney,[13,14] but they have never been interpreted specifically for a biological scientist whose research includes quantal response bioassays. Models in quantal response bioassays range from very simple to extremely complex. The simplest model is a binary model (e.g., dead, alive) with one explanatory variable (e.g., dose). A more complex model is a binary model with multiple explanatory variables; the most complex model is a multinomial (or polytomous) model, in which multiple responses are related to multiple explanatory variables.

Specialized bioassays that screen insects for tolerance to chemicals require far larger sample sizes per dose as well as more doses. These doses must be carefully chosen and placed to obtain credible results. Far-reaching extrapolation should be avoided in any regression analyses. Although the recommendations mentioned in this book are intended to answer very practical problems, the sample sizes that would be necessary to provide valid results are impractical because of the realities of insect rearing or collection from wild populations. The lesson here is that even a well-planned and perfectly executed bioassay with a very simple statistical model cannot answer an impossible question. For example, in commodity treatment for pest exclusion, estimates for a response level (99.9968% mortality) are used although precision defies statistical reality.[15]

Often, pesticide bioassays are done in the highly unnatural environment of a laboratory because arthropod cultures are inexpensive, available year-round, and preliminary research reduces the incidence of unnecessary environmental pollution that would result if every candidate pesticide were tested in a natural environment against a native pest population. How can results from laboratory experiments be used to predict what will actually happen in a field experiment or under the various conditions of operational use?

A large negative with many laboratory bioassays is that too many variables are controlled. Often, these variables include a constant temperature, photoperiod, and relative humidity that promote optimal growth and survivorship. In addition, artificial diets are a convenient and inexpensive means to rear large numbers of insects, but can we assume that their use does not alter the insects' enzyme levels and, hence, their ability to detoxify pesticides? We should not make this assumption without testing to see if there are differences in response between artificial and natural diets. Test subjects are often selected from laboratory colonies so that their ages, sizes, or weights are uniform. Do test subjects selected for uniformity and raised in a controlled environment resemble members of a population in nature? Hardly. Highly controlled conditions are useful if a bioassay is done to identify the effect of an isolated variable on response, but a totally different approach is necessary for realistic prediction.

Perhaps the effects of all variables known to affect response might be summed and a realistic prediction would result. As shown by Robertson and Haverty,[16] this approach involves assigning a weighted relative value to the variable (i.e., a guess about its relative importance) and nothing else. One means of making laboratory pesticides more realistic is to use natural host foliage and apply the pesticide in a way that simulates application in the field. Next, conversion factors that relate laboratory application rates to those in the environment can be derived, as done by Haverty and Robertson.[17] Although this approach is a vast improvement over previous methods, better laboratory bioassays for reliable prediction of arthropod population response in the field can still be developed. What about acceptance that more than one life stage will always be present in field populations, and that control mortality is not a perfect 10%?

Population toxicology,[18] a more ecological approach to the problem of reliable prediction, is discussed. Just as Kogan[19] emphasizes the need for pest management practices in the field to be based on sound ecological theory, pesticide bioassays in the laboratory or greenhouse should be based on principles of population ecology. With this new emphasis, responses of test subjects that simulate a population rather than subjects selected for their uniformity would be evaluated in relationship to physiological time during population development rather than either to chronological time or to isolated variables.

Realistic rearing and test conditions, such as variable temperature regimes, photoperiod, and moisture, can be simulated (e.g., with programmable temperature cabinets). A simulated population might be allowed to develop and be exposed to a pesticide at various intervals during population development.

We present alternatives to the Probit 9 screening for invasive species and discuss more statistically reasonable approaches to these problems. When the probability of death is the only criterion used to estimate the future establishment of a pest population, Q_9^{15} is a general term that can replace automatic use of the probit model.

Once a bioassay is complete, data analyses follow. Numerous techniques for analyses of pesticide bioassay data are described in later chapters, together with cautions about their possible misuses. Many dose–response bioassays provide data that can be analyzed by fairly standard techniques, such as probit or logit analysis, but this is not always the case. Suppose rate of mortality is of interest; probit or logit techniques may or may not be appropriate. If not, what else can be done? Alternative models such as the Weibull or complementary log–log might be more appropriate.[20–22]

Most biologists lack the programming skills necessary to use experimental statistical programs. So, for the sake of statistical mavens everywhere, we primarily illustrate data analyses with two programs that are currently available (i.e., R and Polosuite), useful, and comprehensible. In cases where we illustrate use of an experimental statistical program, we first provide the commands for the program, followed by the analysis itself. Many commercial computer programs are written to do analyses of quantal response data. However, the latter program is the only one that incorporates hypothesis testing into its estimates of probit analysis.

Regardless of which program is used, each analysis must be appropriate for each experiment. Researchers must not succumb to a tendency to force data through the same computer program just because they know how to use that program. Familiarity with one particular program does not guarantee valid analysis. Mindless experimental design that results in data subjected to equally mindless analyses is a waste of time and resources.

Related to the importance of experimental design is the need to avoid untested assumptions that could invalidate an otherwise perfectly designed bioassay. For example, does the response of every arthropod species to a pesticide or any chemical always vary in proportion to body weight? Experimental evidence suggests that sometimes it does, that sometimes it does not, and that no general assumption for either all arthropod species or for all pesticides tested on one species should be made. The effects of other explanatory variables, such as temperature and relative humidity, have been studied with a few pesticides on relatively few arthropod species; the role of these variables in response is also equivocal.

Natural variation has been the most frequently ignored factor in quantal response bioassays. A researcher should not be surprised when the results of repeated bioassays with the same chemicals tested on laboratory population are not exactly the same. Shifts in response over generations have been documented by Savin et al.[23] for western spruce budworm (*Choristoneura occidentalis* Freeman). Such patterns are important, especially if one is tempted to label relatively small differences in response as indicative of resistance to a pesticide. The need for further investigations into topics such as this also is important. We show the equations necessary for estimation of the boundaries of natural variation in response based on large sample theory.

Variation within a population occurs at various levels, including sibling groups, cohorts within a population, and developmental stage. A more ecological approach to the description of population response based on the variability within a population may be possible if, for example, contour plots of mortality[24] in a population over time are integrated with ecologically based methods based on life tables and demographic toxicology.[25]

In bioassays of microbial pesticides, the activity of an unknown preparation is compared with that of a standard. We describe in detail interpretation of potency and relative potency for these pesticides. Because of natural variation, the estimated potency of production lots of *B. thuringiensis* varies even within the same laboratory over time. We also examine resistance screening for tolerance to these microbials.

The evolutionary consequence of pesticide use (and overuse) has been the development of pesticide resistance by numerous important pest species.[26] An entomologist often screens for resistance by comparison of field populations to that of a laboratory population. When wild insects are brought into the laboratory for testing, they are collected from multiple locations where various chemicals have been applied. Does variation among the field-collected groups indicate resistance, variation that is related to the host, or natural variation?

The most reliable criterion for a diagnosis that resistance has developed is failure of a chemical to manage arthropod populations in the field.[27] Quantal response bioassays provide basic information for studies of pesticide resistance. Estimated responses of parental susceptible and resistant strains and their F_1 progeny are often used to test hypotheses about modes of inheritance (e.g., Hoy et al.[28]).

This edition of *Bioassays with Arthropods* will offer a clear explanation of statistical methods to test hypotheses about the genetic mechanisms of resistance based on quantal response bioassays.[29] Two chapters will also explain methodology to handle natural variation in response and offer guidance in the design of experiments to monitor resistance in populations collected from the field.

How can bioassays be designed for this purpose? Mechanical processes and equipment used to do pesticide bioassays have been described in detail in books by Busvine[30] and Shepard.[31] Generally, these aspects of pesticide bioassay have remained unchanged since publication of Busvine's book in 1971. A literature search on bioassays done with either the species of concern or one closely related to it should suffice to provide the necessary details. Regardless, trial and error will be necessary to develop the bioassay to suit the arthropod of interest, laboratory space, research budget, and available personnel.

Thus far, this introduction has concerned bioassays done with a single pesticide. Bioassays with mixtures (pesticide + pesticide, pesticide + microbial agent, or any other combination) have become more frequent during the past few years as resistance to single chemicals have risen, as pesticide costs have increased, and as new ways to decrease environmental pollution have been sought. Although we describe experimental designs for bioassays with mixtures in Chapter 10 (Mixtures), a problem frequently encountered with mixture bioassays occurs not at the level of experimental design, but once the data have been collected. How does one know whether the components are acting independently, synergistically, or antagonistically? Do categories such as "true, actuated, offset, pseudo-synergistic and similar"[32] make any biological sense? We discuss methods to test hypotheses about various kinds of chemical interaction to determine if the combinations of chemicals result in significant differences in efficacy when compared with their individual components. We demonstrate these concepts by describing and showing examples.

Bioassays of response over time require special experimental designs. We explain appropriate designs to use in bioassays that concern time and how to avoid common mistakes in data analyses when time is a variable. Probit or logit techniques may or may not be appropriate. What is a serial experimental design? What is an independent design? What other models might be used to analyze time data? We discuss the use of alternatives such as the complementary log–log model.[33] Again, examples are used to demonstrate how and when to use time as a variable, and sample R code, results, and discussion show real-life outcomes.

Suppose you assumed that a variable has no significant role in response when, in fact, it is crucial. What happens if the appropriate model is not binary but includes two or more explanatory variables? When modules are constructed, they should include all of the relevant explanatory variables. A difficult model would involve multiple explanatory variables and multiple responses. An untested assumption can invalidate an otherwise perfectly designed bioassay. For example, does response to a pesticide or another factor always vary in proportion to body weight? Experimental evidence suggests that sometimes it does, that sometimes it does not, and that no general assumption should be made.[34,35] Many investigators collect information about the multiple effects that occur after treatment with classes of pesticides such as juvenile hormone analogs or benzoyl phenylureas but then do not know how to analyze their data. In this book, we explain how to use these models and how to analyze such data.

In the future, it will be important to transition from bioassays with laboratory populations to toxicology at the levels of populations, communities, and the ecosystem. Statistical models and ecological models are not the same. The former use experimental data (the more, the better) to describe mathematical relationships between and among variables. Ecological models, in contrast, are conceptual maps about components of a particular ecosystem.[36] Such models tend to include simplified descriptions of the various components, along with weighting factors for their importance. Ecotoxicology has been defined as the science of predicting the effects of toxicants on ecosystems and on nontarget species[37]; the effects of toxicants at the individual, population, and ecosystem level are of concern.[38] How can quantal response data for arthropods be used in predictive, conceptual models?

In the final chapter of this book, we suggest some possible steps to identify all significant explanatory variables and multiple response variables that should be included in a comprehensive multinomial approach to ecotoxicology. A major improvement necessary to make quantal response bioassays more predictive is the use of response variables other than simply "dead versus alive"[39] and to use statistical methods especially for prediction. Stark and Wennergren[40] and Stark and Banks[41] described modifications to the demographic approach to toxicology, but only one life table parameter was considered.

REFERENCES

1. Anonymous, Insect sting allergy. American College of Allergy, Asthma, & Immunology. http://acaai.org/allergies/types/insect-sting-allergies. 2016.

2. California Department of Food and Agriculture, *Draft Report of Environmental Assessment of Pesticides Program*, Vol. 3, CDFA, Sacramento, CA, 1978, 5.94-1.

3. Mayr, E., *Principles of Systematic Zoology*, McGraw-Hill, New York, 1969.

4. Shipp, J. L. and Zhang, Y., Using greenhouse microclimate to improve the efficacy of insecticide application for *Frankliniella occidentalis* (Thysanoptera: Thripidae), *J. Econ. Entomol.* 92, 201, 1999.

5. Lester, P. J. and Barrington, A. M., Gamma irradiation for post-harvest disinfestation of *Ctenopseustis obliquana* (Walker) (Lepidoptera, Tortricidae), *J. Appl. Entomol.* 121, 107, 1997.

6. Jang, E. B., Nagata, J. T., Chan, H. T., Jr., and Laidlaw, W. G., Thermal death kinetics in eggs and larvae of *Bactrocera latifrons* (Diptera: Tephritidae) and comparative thermotolerance to other tephritid fruit fly species in Hawaii, *J. Econ. Entomol.* 92, 654, 1999.

7. Roberts, T., *100% Organic Pest Control for Home & Garden*, Book Publishing Company, Summertown, TN, 1998, and Ellis, B. W. and Bradley, F. M., eds., *The Organic Gardener's Handbook of Natural Insect and Disease Control*, Rodale Press, Emmaus, PA, 1996.

8. Beegle, C. C. and Yamamoto, T., Invitation paper (C. P. Alexander Fund): History of *Bacillus thuringiensis* research and development, *Can. Entomol.* 124, 587, 1992.

9. Shapiro, M. and Robertson, J. L., Natural variability of three geographic isolates of gypsy moth (Lepidoptera: Lymantriidae) nuclear polyhedrosis virus, *J. Econ. Entomol.* 84, 71, 1991.

10. Anderson, T. E., Hajek, A. E., Roberts, D. W., Preisler, H. K., and Robertson, J. L., Colorado potato beetle (Coleoptera: Chrysomelidae): Effects of combinations of *Beauveria bassiana* with insecticides, *J. Econ. Entomol.* 82, 83, 1989.

11. Burges, H. D., *Microbial Control of Pests and Plant Diseases, 1970–1980*, Academic Press, London, 1981.

12. van den Bosch, R., *The Pesticide Conspiracy*, Doubleday, Garden City, NY, 1976.

13. Finney, D. J., *Probit Analysis*, Cambridge University Press, Cambridge, England, 1971.

14. Finney, D. J., *Statistical Method in Biological Assay*, Griffin, London, 1964.

15. Robertson, J. L., Preisler, H. K., and Frampton, E. R., Statistical concept and minimum threshold concept, in Paull, R. E. and Armstrong, J. W. eds., *Insect Pests and Fresh Horticultural Products: Treatments and Responses, CAB* International, Wallingford, England, 1994, pp. 47–65.

16. Robertson, J. L. and Haverty, M. I., Multiphase laboratory bioassays to select chemicals for field-testing on the western spruce budworm, *J. Econ. Entomol.* 74, 148, 1981.

17. Haverty, M. I. and Robertson, J. L., Laboratory bioassays for selecting candidate insecticides and application rates for field tests on the western spruce budworm, *J. Econ. Entomol.* 75, 179, 1982.

18. Robertson, J. L. and Worner, S. P., Population toxicology: Suggestions for more predictive bioassays, *J. Econ. Entomol.* 83, 8, 1990.

19. Kogan, J., ed. *Ecological Theory and Integrated Pest Management Practice*, John Wiley & Sons, New York, 1986.

20. Dell, T. R., Robertson, J. L., and Haverty, M. I., Estimation of cumulative change of state with the Weibull function, *Bull. Entomol. Soc. Am.* 29, 38, 1983.

21. Preisler, H. K. and Robertson, J. L., Analysis of time-dose-mortality data, *J. Econ. Entomol.* 82, 1534, 1989.

22. Copenhaver, T. W. and Mielke, P. W., Quantit analysis: A quantal assay refinement, *Biometrics,* 33, 175, 1977.

23. Savin, N. E., Robertson, J. L., and Russell, R. M., A critical evaluation of bioassay in insecticide research: Likelihood ratio tests of dose-mortality regression, *Bull. Entomol. Soc. Am.* 23, 257, 1977.

24. Robertson, J. L., Richmond, C. E., and Preisler, H. K., Lethal and sublethal effects of avermectin B_1 on the western spruce budworm (Lepidoptera: Tortricidae), *J. Econ. Entomol.* 78, 1129, 1985.

25. Stark, J. D. and Banks, J. E., Population-level effects of pesticides and other toxicants on arthropods, *Ann. Rev. Entomol.* 48, 505, 2003.

26. Roush, R. T. and Tabashnik, B. E., eds., *Pesticide Resistance in Arthropods*, Chapman & Hall, London, 1990.

27. Ball, H. J., Insecticide resistance: A practical assessment, *Bull. Entomol. Soc. Am.* 27, 261, 1981.

28. Hoy, M. A., Conley, J., and Robinson, W., Cyhexatin and fenbutin-oxide resistance in Pacific spider mite (Acari: Tetranychidae): Stability and mode of inheritance, *J. Econ. Entomol.* 82, 11, 1988.

29. Preisler, H. P., Hoy, M. A., and Robertson, J. L., Statistical analysis of modes of inheritance for pesticide resistance, *J. Econ. Entomol.* 83, 1649, 1990.

30. Busvine, J. R., *Critical Review of the Techniques for Testing Insecticides*, Commonwealth Agricultural Bureau, London, 1971.

31. Shepard, H. H., ed., *Methods of Testing Chemicals on Insects, II*, Burgess Publishing, Minneapolis, MN, 1960.

32. Sakai, S., Joint action of insecticides, *Bull. Daito Bunko Univ.* 1, 1, 1969.

33. Preisler, H. K. and Robertson, J. L., Analysis of time–dose–mortality data, *J. Econ. Entomol.* 82, 1534, 1989.

34. Robertson, J. L., Savin, N. E., and Russell, R. M., Weight as a variable in the response of western spruce budworm to insecticides, *J. Econ. Entomol.* 74(5), 643, 1981.

35. Savin, N. E., Robertson, J. L., and Russell, R. M., The effect of body weight on lethal dose estimates for the western spruce budworm, *J. Econ. Entomol.* 75(3), 538, 1982.

36. Jorgensen, S. E. and Bendoricchio, G., *Fundamentals of Ecological Modelling*, Elsevier, Amsterdam, 2001.

37. Hoffman, D. J., Rattner, B. A., Burton, G. A. Jr., and Cairns, J. Jr., Introduction, in D. J. Hoffman, B. A. Rattner, G. A. Burton, Jr., and J. Cairns, Jr., eds., *Handbook of Ecotoxicology*, Lewis Publishers, Boca Raton, FL, 2003, pp. 1–16.

38. Kammenga, J. and Laskowski, R., Demographic approaches in ecotoxicology: State of the art, in Kammegna, J. and Laskowski, R., eds., *Demography in Ecotoxicology*, John Wiley & Sons, New York, 2000, pp. 3–8.

39. Robertson, J. L. and Worner, S. P., Population toxicology: Suggestions for laboratory bioassays to predict pesticide efficacy, *J. Econ. Entomol.* 83, 8, 1990.

40. Stark, J. D. and Wennergren, U., Can population effects of pesticides be predicted from demographic toxicological studies?, *J. Econ. Entomol.* 88, 1089, 1995.

41. Stark, J. D. and Banks, J. E., Population-level effects of pesticides and other toxicants on arthropods, *Annu. Rev. Entomol.* 48, 505, 2003.

CHAPTER 2

Quantal Response Bioassays

Dr. Maven is well aware that her first attempt at doing a bioassay was a disaster. "Obviously," she reasons, "I need some help." And so, she enrolls in a one-day workshop given at the Griffindor University Department of Entomology. Given her status as a faculty member who does not want to reveal her ignorance, she brings along her research associate Jessica Flapperjack to pose as the student in much need of education in bioassays. Dr. Maven comes along to the meeting to just offer moral support.

The air is warm and still as nine students set off for bioassay lessons. Their instructor is Dr. Garland Tarleton, a kind, middle-aged, portly fellow with an Albert Einstein coiffure and a lively sense of humor. After he divides his nine students into groups of three, Dr. Tarleton gives each group the choice of a 0.357 magnum or 0.44 magnum water pistol each with a bucket of water (i.e., the single chemical) or a precision paintball gun with a supply of red paintballs. Dr. Tarleton warns the groups to choose their weapon carefully because he cannot allow them to change midexperiment. Group 1 selects the paintball gun; groups 2 and 3 each choose a water pistol.

The target for each group differs. Group 1's target is an electronically controlled model *Larsonesquc corcid* (leaf-footed bug[1]) with sensors; when paint contacts the bug's body, the model flips over and plays dead. Group 2's target is a very small circle with eight legs (possibly a grossly magnified spider mite) drawn on a sheet of acetate. Group 3 also has an acetate target, but the drawing is a large outline of a caterpillar's head. Each group receives the same instruction: In a replicated experiment, estimate the distance for which the probability of hitting the target is 50%.

Let's compare determining the range of doses for a bioassay with that of the distance of each student to their targets. How do the students begin? Their first task should be to locate the distance beyond which the target is out of range. The second step is to find the distance at which it is virtually impossible to miss. Everyone agrees that these boundaries or limits need not be defined with utmost precision and that the real experiment (within these boundaries) should be the focus of most effort.

A major problem arises when groups 2 and 3 try to define a hit. In the case of a bioassay, this hit would be referred to as the particular response you wish to measure. Group 1 is not concerned with this problem because their model target just flips over and plays dead. There is no gray area for how to measure a success.

Groups 2 and 3, however, must define a hit more arbitrarily. Group 2 decides to count any contact of water with the body or legs of the spider mite as a hit. The members of group 3 decide that they can hit their caterpillar in two ways, on the lines and within the lines.

Groups 2 and 3 move away from group 1 because they are tired of being hit by stray paintball bullets, so they move into a convenient canyon where the wind is blowing in intermittent gusts. As they begin their experiment, it becomes clear to each group that the wind affects their accuracy. One member of each group is trusted with a top-of-the-line anemometer to record wind speed. Jessica Flapperjack is promptly handed the small gadget because no one else in her group can be held responsible to not lose it. Both groups record distance from the target and wind speed for each shot. The enthusiasm by the members finally able to shoot their target is quickly dampened when they realize they cannot easily see the water on their targets. They look up in frustration to Dr. Tarleton, who shifts his smile to a smirk. Ok, the groups will make their current method work.

The next problem concerns the definition of a replication. Should one person do all the shooting? If not, in what order should group members take their turns? Group 1 assigns the order in which members shoot the paintball gun at random and to let everyone take one shot before any one person shoots again. Group 2 independently comes to the same decision. Group 3, however, decides to let one person do all the shooting, another person serves as the target observer, and another individual records wind speed when each shot is taken.

The groups have determined their limits and what quantifies as success (100%), but perhaps they wish to know what criteria are necessary to have only 50% success rate. As the experiments proceed, no one has any idea how many shots of water pistol are necessary to estimate the distance from the target where the probability of hitting the target is 50%. Certainly, each group should be able to give a rough estimate, but how reliable will it be? Is each group absolutely sure that they will obtain the same 50% distance if they repeated the experiment under exactly the same conditions? Is there an absolute, true answer to this problem? Furthermore, how can group members estimate a probability with these kinds of data? Will group 1 ever get the paint out of their clothes? Will groups 2 and 3 ever dry off?

Dr. Maven's notes from the course provide a good introduction to the general subject of bioassays. After the experiments, the instructor provides the following general information and critique.

2.1 TYPES OF QUANTAL RESPONSE BIOASSAYS

These experiments illustrate a *quantal response* bioassay, an experiment in which the response of the test organisms varies in action to a measurable characteristic of a stimulus. In the water pistol experiment, the target represents the arthropod being tested. The paintball gun or water pistol represents the application device for the treatment. Just as data collected by the three groups differ, quantal response in bioassays can differ in terms of the *response variables* (dependent variables) and the *explanatory variables* (independent variables). Explanatory variables are the measurable characteristics of a stimulus or stimuli that cause a response or responses of the test organism to vary. Response variables are the random outcomes of the experiment; they vary in relation to the stimulus involved.

Group 1 did the simplest type of quantal response experiment. It dealt with only one explanatory variable (distance from the target) to estimate an all-or-nothing result (flip over dead or stay as is). This kind of test is a binary quantal response experiment with one explanatory variable. Group 2 measured two explanatory variables (distance from the target and wind speed), but its members also recorded an all-or-nothing result (hitting the spider mite or missing it). This type of test illustrates a binary quantal response experiment with multiple explanatory variables. Members of group 3 attempted to do the most complex type of experiment. They measured two explanatory variables (distance from the target and wind speed) and two response variables (hits on the lines and hits within the lines). Theirs was a polytomous (multinomial) experiment.

The frequency with which these different types of experiments have been used in pesticide research in biological experiments seems to vary as a direct function of the complexity of the data analyses involved. Because it is the simplest, the binary quantal response experiment with one explanatory variable (dose–response or concentration–response) is done most frequently. Binary experiments with multiple explanatory variables (i.e., quantal response (multivariate) experiment) are not done very often, perhaps because they are more time-consuming and because the statistical methods used to analyze the data are complex.[2,3] The polytomous quantal response experiment is extremely time-consuming and too complex to be utilized by any scientist without a hefty statistics background until recently. Previously, reliable computer programs written specifically for polytomous quantal response experiments were not available except for analyses of very small data sets.

2.2 EXPERIMENTAL DESIGN OF BIOASSAYS

The way a bioassay is done is its *experimental design*. Four general aspects of experimental design are common to all quantal response bioassays. First, the design must describe the *experimental unit*—the entity actually receiving the treatment.[4] Examples of experimental units are single insects (e.g., Li and Otvos[5]) or cages containing plants or fruit.[6,7] Treatments frequently involve use of groups, rather than individuals, exposed to each treatment level. For some species, a standard procedure is to select individuals from a laboratory colony, place them in groups of 10 into a Petri dish or other container, and treat each insect with a measured drop (e.g., Robertson and Kimball[8]). In this instance, the individual insect is the experimental unit. Suppose that the entire group is sprayed simultaneously or presented with treated diet. In this situation, the group (not each insect) is the experimental unit (e.g., Robertson and Boelter[9]).

In a quantal response bioassay, nothing except the treatment level or intensity can distinguish one group from another: the experimental unit must be carefully defined or identified before the bioassay begins. Occasionally, an unforeseen event (such as cannibalism or a simple mistake in counting) can occur without affecting the integrity of the experiment. Numbers treated are just adjusted after the bioassays are completed.

In experiments to select arthropods for resistance to a treatment, a preplanned minor modification of the initial experimental unit (for example, from use of two insects in each container) to another unit (for example, use of one insect in solitary confinement) may also be necessary as the insects grow and develop (e.g., Robertson and Kimball[8]).

2.2.1 Randomization

Once the group has been defined (e.g., larvae in a particular developmental stage might be the target population), test subjects must be selected from all available individuals in the group as randomly as possible. Random selection is the only way to avoid experimental bias, so that results adequately represent responses of the target group. The process of assigning experimental units to treatments and controls at random is known as *randomization*.

The replication process usually involves groups of test subjects (rather than individuals) treated with each pesticide dose or concentration. Regardless of the type of quantal response bioassay, the groups should be assigned randomly to each dose or concentration in a manner similar to that used. To ensure randomization, use of a random number table or online randomization program is advisable.

If selection of experimental units is not randomized, for example, large insects might be selected preferentially, as might larvae that move most slowly. Data, once analyzed, would then pertain to subsets of the population (large or sluggish caterpillars) and not to the population as a whole.

Suppose, for example, that comparative responses of last instars (the target or test population) of a particular species of Lepidoptera to several pesticides are of interest. The target group in this case is composed of all individuals of the last instar that are available for testing. Now suppose that this species is being reared in laboratory culture, and 100 Petri dishes contain insects in the target instar. One way to randomly select test subjects is to number the Petri dishes from 1 to 100. For one replication, 10 Petri dishes might contain sufficient numbers for testing. The 10 Petri dishes to be used in a given replication could be those corresponding with the first 10 numbers between 1 and 100 in a list of random numbers.

Logistical convenience can result in bioassays with invalid designs if randomization does not rectify the issue.[4] In the laboratory or greenhouse, for example, insects are often held in stacks of Petri dishes or rearing containers placed on shelves where gradients in light intensity and temperature might significantly affect response. A sample consisting only of insects selected from the top dishes at the front of shelves is not a random sample of the population. Instead, containers from which individuals are removed should be randomly selected to account for location on shelves and position within stacks.

When wild populations are tested, some modifications of the individualized randomization procedures suited to each population are usually necessary. For example, natural units such as leaves, branches, or cones might be numbered, and insects on each unit can then be selected at random for testing. If the arthropod is to remain in or on the natural unit during testing, then these numbered units can be randomly assigned to treatment levels.

2.2.2 Treatments

In a quantal response bioassay with a single explanatory variable, only the intensity of treatment can be subject to change. Thus, group 1 could change the distance to the target, but not the fact that they were shooting paintballs; groups 2 and 3 were stuck with using their water pistols. When multiple explanatory variables are investigated, the identity of variables also cannot be changed in the midst of the bioassay. Group 2 could change the distance to the target and continue their measurements of wind speed but not change the fact that they were shooting water at the target. Finally, the treatments in a quantal response bioassay with multiple explanatory variables and multiple response variables cannot be changed midway through the experiment: group 3 could not change the variables it measured or the effects it recorded.

Why are these criteria written in stone? When an investigator changes the exposure method (that is, the treatment) midway through an experiment with one explanatory variable, the two (or more) exposure methods become additional explanatory variables. Suppose an investigator also changes the time(s) at which an effect (such as mortality) is assessed midway through the experiment. Without knowing the relationship between time and mortality, versus exposure method, Dr. Tarleton describes the subsequent bioassay as a fruitcake: "A mixture of apples, oranges, walnuts, those strange candied fruits, and flour; who knows what you have? Try to add the results back together into a publishable form: I dare you!" Unfortunately, results of seemingly invalid bioassays occasionally appear in the published literature (see Table 1 in Bolin et al.[10]).

2.2.3 Controls

A control group should be exposed to everything *except* the treatment and must be included in each replication of a bioassay. The rationale for control groups is simple: without them, it will be difficult to attribute an observed effect or response to the treatment with any certainty to the pesticide. For example, a solvent or an impurity in a solvent might be responsible for a particular result. The control group should be exposed to everything to which the subjects are exposed except the pesticide. For example, if the test subjects are exposed to a chemical in water, the control group should be treated only with water.

Likewise to selection of treated units, controls must represent a random sample selected from the target population by the same criteria and standards used to select the test subjects.

Use of a common control group for several treatments in a quantal response bioassay is not desirable even if the same solvent is used. Each treatment tested in a quantal response bioassay must have its own control. To illustrate what might happen if a common control is used, suppose a control group is treated first, followed by groups treated with several levels of treatment A, several levels of treatment B, and several levels of treatment C. None of the controls respond, the test subjects treated with treatment A respond in relation to the dose applied, and all test subjects treated with treatments B and C respond regardless of dose. What does this bioassay reveal about the effects of treatments B and C? The answer is, unfortunately, not very much. Possibly, these two treatments are very effective, and lower rates should be tested. It is equally possible that the application device became contaminated with the highest level of treatment A, and as a result, treatment B became contaminated, followed by contamination of treatment C with treatment B. Given the choice of testing more treatments at the expense of testing a control with each one, the wisest option is to test fewer treatments.

Is an "untreated" control a real control? The answer is no, unless the bioassay is done with the chemical in its pure form (without a carrier or solvent) or unless the fumigation activity of a chemical is being tested. The control treatment must be done with everything except the pesticide. In tests with fumigants, air is the carrier and an untreated control is indeed appropriate. However, when a treatment is applied as an emulsion in water, the appropriate control is water containing all of the emulsifying agents in the undiluted pesticide formulation, then diluted to the same concentrations present in the aliquots actually tested.

2.2.4 Replication

Dr. Garland Tarleton next discusses a common tendency among scientists never to replicate a successful experiment (defined as one in which the results are exactly as expected the first time). Reports of unreplicated bioassays occasionally appear in the published literature (see Table 1 in Bolin et al.[10]). Without exception, all experiments *must* be replicated. A *replication* is repetition of a bioassay both at a given time and at different time (time, but under the same conditions [as much as possible] as the first test). The chemical treatments should likewise be new (i.e., solutions not older than 1 week) and not reused for a replicated experiment. One purpose of replication is to randomize effects related to uncontrollable laboratory procedures and conditions so that experimental error can be estimated. Experimental error includes effects caused by unexplained correlations among the subjects in the same treatment, as well as unexplained variation that occurs each time an experiment is done. A true estimate of experimental error can only be by testing more than one treatment group with the same dose in the same experiment and repeating the experiment at least three times more.

Within the same experiment, sources of correlation may be genetic; that is a group of test subjects in the same Petri dish may be more related by a common ancestor than are other test subjects in another Petri dish. As a result, their responses may be naturally more similar than the responses of test subjects in another Petri dish. In bioassays in which the pesticide has been incorporated into an aliquot of artificial diet upon which all test subjects in a container feed or in which pesticide has been deposited on natural host foliage upon which the test subjects feed and walk, responses are likely to be more similar than responses of insects fed and held individually.[11] Likewise, some conditions under which insects are held after treatment seem to facilitate regurgitation and reflex bleeding.[12,13] Cross-contamination among all of the insects in a Petri dish is an obvious cause of correlated response.

Among identical experiments done at different times, such factors as subtle differences in application techniques, time of day, or small fluctuations in temperature may be sources of experimental error. Replication is necessary to estimate the extent to which these factors cause differences in response time. Another reason for replication is to detect errors in formulation (i.e., in the preparation of the solutions actually tested); for this reason, the best way to ensure true replication is to prepare a fresh series of test concentration doses diluted from a new stock solution each time the experiment is done.

Some investigators refer to subsets within a replication as replications. An example is each group of 10 insects in a Petri dish treated with the same concentration removed from the same container during the same treatment period, or 20 insects might be treated in groups of 10 with concentrations of 0 (the control), 1, 2, 5, 7, and 10 mg/ml of chemical X. Each concentration has been prepared once as one of the dilutions from the same stock solution. For this hypothetical bioassay, each group of 20 treated with the same concentration removed from the same container is a replicate and each of the groups of 10 treated with that concentration is a subset of that replicate. When the experiment is done once and when the subset is called a replicate, the investigator is guilty of *pseudo-replication*.

The problem of pseudo-replication is that the only variation that can be estimated in this situation is the variation in response among insects in the Petri dishes. An estimate of this variation does not include variation that results from differences in the formulation procedure itself (i.e., variation in diluting from the stock solution). Unless each group of 10 treated with any concentration was treated with a fresh formulation and (to be absolutely sure) at a different time, it is *not* a separate replicate. Pseudo-replication is not unique to quantal response bioassays, but it occurs with alarming frequency.

Ideally, replications should be done on different days within a relatively short period of time (e.g., a week). Day-to-day variability must be considered as a possible source of experimental error in the experiment when data are analyzed. This variability is especially important when conducting bioassays on field collected populations to screen for tolerance to chemicals. The longer the time period between screens, the higher the increase in variability between results. Differences in response that occur during the replication process may be a source of extrabinomial variability, the significance of which should be tested. Time constraints related to the condition of test subjects, use of the application equipment, or varying personnel may require modifications of this ideal method of replication.

Decisions about the minimum time that should pass between replications must be made by individual investigators based on their common sense. Just as no specific guidelines are available to define the lower time limit for reliable replication, none are available to identify the maximum time that can pass after which a treatment becomes an altogether different experiment rather than a replicate. Certainly, the same bioassay done with the same species, but in different years, cannot be considered a replication in the usual sense. Because of these problems, administration of a series of fresh solutions or treatment levels is more specific than time as a criterion for true replication.

Suppose that chemical A is to be tested on a group of insects collected from the field, and a shipment of another group from a different site is expected to arrive the next day. Obviously, the time available to do the bioassay is limited. One set of treatment solutions is prepared, but three different groups will be tested one hour apart. Is this true replication? The answer is no, not because only an hour separates the experiments, but because a possible error in preparation of the test solutions cannot be detected. If the experiments were done with fresh solutions, the bioassay would be truly replicated.

Now suppose that fresh solutions are used. In the first replication, the pesticide concentrations that are applied kill between 5% and 95% of the insects. In the second, all of the controls live, but all of the insects treated with the chemical die regardless of concentration applied. Although these results may be valid, a procedural error may have occurred, but in which replicate? Results of the third and fourth replications might suggest that a formulation error occurred in one of the first two replicates. Results from a replicate can be disregarded if (and only if) an error is known to have occurred in that replicate. Data must *never* be discarded without strong justification, such as a known procedural error or an outlier that is detected as part of the data analysis (see Section 3.2.1.1). If an error does not seem likely (e.g., if the results of the third replicate are like those of the first, but the results of the fourth replicate are similar to those of the second), all of the data should be used for statistical analyses. Inconsistency from replicate to replicate may be the result of extrabinomial variability of outliers, and various other factors (see Section 3.2.1.1).

2.2.5 Order of Treatments within a Replication

On both practical and theoretical grounds, Dr. Tarleton suggests that the order in which concentrations of a given treatment rates are applied in a quantal response bioassay should be from lowest concentration to the highest, never vice versa. If the treatment is highest to lowest concentration in a bioassay with a chemical, for example, results are likely to be spurious because the test subjects may actually be exposed to a higher concentration rate than intended. The impracticality of cleaning an application device and letting it dry completely between applications of different concentrations is obvious. However, such cleanings are necessary if concentrations are applied in random order. (Equally impractical is the use of differently calibrated syringes or spray reservoirs for each concentration.) Theoretically, some residual solvent will remain in the application device after cleaning without subsequent drying. If the solvent is pure (as would be the case if cleaning had just occurred), the concentration of the chemical tested next will be diluted more than if solvent plus a lower concentration of chemical remains in the needle, syringe, spray reservoir, or nozzle. In the treatment sequence of lowest to highest concentration rate, the application device need not be cleaned.

When more than one treatment is tested in a bioassay as part of each replication, the order in which the chemicals are applied should be randomized among replications. Suppose that, unknown to the investigator, small amounts of chemical A interact with chemical B to cause more mortality and have a greater effect than would be expected if the joint effects of the two chemicals were independent and additive. Each time chemical B was tested, the effects that are observed will not just be due to B alone if it is always tested after pesticide A. Randomization of the pesticide treatment sequence of chemicals helps minimize the possible experimental bias and error that may occur.

2.3 COMPUTER PROGRAMS

Computer programs are not only crucial for analyses of bioassay data, but also, their manuals and documentation often provide a useful guide to the statistical literature. According to Dr. Tarleton, before the availability of personal computers and access to large statistical packages that once ran only on mainframe computers, analyses of bioassay data were a horrendous chore. Programs developed by individual scientists for specific methods were occasionally developed for programmable calculators (e.g., Toth and Sparks[14]) but their documentation tended to be vague. FORTRAN code for special procedures[15] was published by some authors but was rarely, if ever, used by other investigators; these programs and routines were usually used exclusively by the people who developed them. Large commercial statistical packages sometimes do not automatically provide enough information to prevent the user from doing the wrong test on the wrong type of data and then reaching the wrong conclusions.[16] Whatever program is used, it must be capable of producing all of the necessary test statistics so that maximum information can be extracted from the bioassay data, and all hypotheses of interest can be tested.

Dr. Tarleton concludes the bioassay workshop with sample data sets from binary quantal response bioassays with one or more explanatory variables,[17,18] along with their analyses with two user-friendly computer programs. These programs were designed specifically to provide biologists bumbling their way through bioassays from going too far astray. In addition, they are computationally correct and well documented. As a contrast, he also provides command lines for analyses of selected data sets with the statistical-computing environment R.[19] He recommends that a user-friendly program should be used if available. If not, a well-documented statistical package[17,18] can be used with proper guidance.

REFERENCES

1. Mitchell, P. L., Leaf-footed bugs (Coreidae), in Schaefer, C. W. and Panizzi, A. R., eds., *Heteroptera of Economic Importance*, CRC Press, Boca Raton, FL, 2000, pp. 337–403.
2. McCullagh, P. and Nelder, J. A., *Generalized Linear Models*, CRC Press, Boca Raton, FL, 1999.

3. Colin, D., Models for teratological data, in Krewski, D. and Franklin, C., eds., *Statistics in Toxicology*, Gordon and Breach Science Publishers, New York, 1991, pp. 335–347.

4. Krewski, D. and Bickis, M., Statistical issues in toxicological research, in Krewski, D. and Franklin, C., eds., *Statistics in Toxicology*, Gordon and Breach Science Publishers, New York, 1991, pp. 11–41.

5. Li, S. Y. and Otvos, I. S., Comparison of the activity enhancement of a baculovirus by optical brighteners against laboratory and field strains of *Choristoneura occidentalis* (Lepidoptera: Tortricidae), *J. Econ. Entomol.* 92, 534, 1999.

6. Stark, J. D. and Wennergren, U., Can population effects of pesticides be predicted from demographic toxicological studies?, *J. Econ. Entomol.* 88, 1089, 1995.

7. Bergh, J. C., Rugg, D., Jansson, R. K., McCoy, C. W., and Robertson, J. L., Monitoring the susceptibility of citrus rust mite (Acari: Eriophyidae) populations to abamectin, *J. Econ. Entomol.* 92, 781, 1999.

8. Robertson, J. L. and Kimball, R. A., Toxicities of topically applied insecticides to western spruce budworm, 1979, *Insecticide Acaricide Tests* 5, 203, 1980.

9. Robertson, J. L. and Boelter, L. M., Toxicity of insecticides to Douglas-fir tussock moth, *Orgyia pseudotsugata* (Lepidoptera: Lymantriidae) I. Contact and feeding toxicity, *Can. Entomol.* 111, 1145, 1979.

10. Bolin, P. C., Hutchison, W. D., and Andow, D. A., Long-term selection for resistance to *Bacillus thuringiensis* Cry1Ac endotoxin in a Minnesota population of European corn borer (Lepidoptera: Crambidae), *J. Econ. Entomol.* 92, 1021, 1999.

11. Preisler, H. K. and Robertson, J. L., Analysis of time–dose–mortality data, *J. Econ. Entomol.* 82, 1534, 1989.

12. Lang, J., Effects of regurgitation and reflex bleeding on mortality in western spruce budworm (*Chloristoneura occidentalis* Freeman) treated with Iannate, *Entomol. Exp. Appl.* 12, 288, 1969.

13. Robertson, J. L. and Lyon, R. L., Douglas-fir tussock moth: Contact toxicity of 20 insecticides applied to the larvae, *J. Econ. Entomol.* 66, 1255, 1973.

14. Toth, S. J. Jr. and Sparks, T. C., Effect of temperature on toxicity and knockdown activity of *cis*-permethrin, esfenvalerate, and λ-cyhalothrin in the cabbage looper (Lepidoptera: Noctuidae), *J. Econ. Entomol.* 83, 342, 1990.

15. Robertson, J. L. and Smith, K. C., MIX: A computer program to evaluate interaction between chemicals, *USDA Forest Service Gen. Tech. Rep.*, PSW-112, 1989.

16. Gad, S. G., *Statistics and Experimental Design for Toxicologists*, CRC Press, Boca Raton, FL, 1998.

17. LeOra Software, *PoloJR*, LeOra Software LLC, PO Box 563, Parma, MO 63870. 2016. See http://www.LeOra-Software.com.

18. LeOra Software, *PoloMulti*: Multiple probit or logit analysis for Windows and OS, LeOra Software, PO Box 563, Parma, MO 63870. 2016. See http://www.LeOra -Software.com.

19. R version 1.9.1, released on June 21, 2004. See http://www.r-project.org.

CHAPTER **3**

Binary Quantal Response
with One Explanatory Variable

By training, Dr. Maven is an insect ecologist. In graduate school, she was proud to be a student of "pure" science. She did not consider applied problems to be worthy of her attention. But her move to the Schaeferville Experiment Station has changed her perspective. Now she is responsible for developing the Research Station's pesticide bioassay program and saving Schaeferville from *Patronius giganticus*. This chapter and those to follow chronicle Dr. Maven's education in the basic principles of pesticide bioassay. She learns many lessons by trial and error; the problems that she encounters are described here. The rest of her expertise is gained by reading relevant references on the subject, many of which are cited in this book. Procedures for data analyses for each kind of quantal response bioassay are thoroughly explained.

The inherent advantages and limitations of each statistical method are described, and reliable computer programs that are available for each kind of data analysis are discussed. Output from some of these programs is shown in detail to illustrate biological interpretation, but use of these programs as examples does not mean that they are perfect or that they will not be replaced by better programs in the future. The need for further improvements becomes apparent to Dr. Maven as her bioassays become more sophisticated and she needs more and more special information from her experimental results.

As mentioned in Chapter 2, the simplest quantal response bioassay is a binary response experiment with one experimental variable. This is the kind of bioassay that Dr. Maven does first with the laboratory colony of the Godfather worm (*P. giganticus*) that she has finally established. Her graduate research concerned development of a computer model to predict outbreaks of the Godfather worm on broom (*Cytisus scoparius* L.). She concluded that peak population levels occur on March 24 of each calendar year; moth flights begin at dusk and continue until midnight on that night in particular. Because of the close systematic relationship between the *black witch* (*Ascalapha odorata* L.) and the Godfather, Paula originally thinks that the new species can be studied easily. She is so very wrong.

First, the wings of adult *P. giganticus* are covered with scales that incessantly shed with each wing flap to the point that the laboratory resembles a dust cloud in the Arizona desert. Second, the saliva of larvae contains an exotic, ghastly, and unidentified chemical that renders fruit and other plant material totally inedible. Third, tomatoes sent to processing plants explode at random and wreck expensive machinery. Finally, larvae preferentially chew near xylem, which spreads the chemical throughout the plant tissue very rapidly. Laboratory rearing might prove to be a bit complicated.

Before anything else can be done, Paula Maven's technician, Jessica Flapperjack, requisitions face masks and a top-of-the-line air cleaner for the lab. Then, based on simple logic, common sense, and familiarity with every insect diet ever developed,[1,2] Jessica develops a new one specifically for this insect. She combines tomato paste, garlic, and onions in a large blender; suspends the mixture in autoclaved, molten agar; then pours the cooling tomato-agar mixture onto diet trays. The insects eat the diet hungrily and the laboratory population soon flourishes. Larvae reach about 6 cm in length when mature. Only 30 days pass from the egg to the adult stage. Now the bioassays can begin.

The simplest quantal response bioassay is a binary quantal response experiment with one explanatory variable. This is the kind of bioassay that Dr. Maven has Jessica do first with the laboratory colony of *P. giganticus*. The bioassays are done for three reasons. First, Paula needs to estimate the relationship between dose and response and compare the regression lines for individual chemicals. For each generation of laboratory rearing, Paula uses three types of chemicals—organophosphate, pyrethrin, and carbamate—as standards to monitor changes in response across generations. Second, she must compare the doses that cause the same levels of mortality. For these experiments, Jessica tests 10–15 toxicants to estimate their toxicities relative to responses to the three standard chemicals tested in the same generation. Third, Jessica tests several geographically separate strains of nuclear polyhedrosis virus on the Godfather worms. For these experiments, Dr. Maven will identify the groups of strains with equal response. Thus, three types of comparisons—among lines, among points on lines, and among groups of lines—are necessary.

3.1 TERMINOLOGY AND GENERAL STATISTICAL MODEL

Any binary response bioassay typically involves the selection of several doses, administration of each dose to a number of test subjects, and, after a specified time, designation of each subject's response as either yes (e.g., dead, knocked down) or no (e.g., alive, not knocked down). In arthropod toxicology, the definition of *dose* has become unnecessarily complicated.[3,4] How does a *dose* differ from a *concentration*? Many investigators use the terms interchangeably, but a dose is properly defined as the amount of a chemical applied per unit weight of the test subject. This definition resembles that of Finney as "an intensity specified in units of concentration, weight, time, or other appropriate measure…"[5] Similarly, Hamilton[6] has defined dose as a generic term for the level of exposure. We will use Finney's[5] definition here.

The statistical statement of the binary quantal response with a single explanatory variable is shown in Equation 3.1:

$$P_i = F(\alpha + \beta x_i), \tag{3.1}$$

where P_i is the probability of response, x_i is the ith dose or a function of that dose (e.g., logarithm of dose), and F is a distribution function, and $\alpha + \beta x_i$ is the regression line. Why is this statistical relationship used rather than a simple estimation of the relationship from a graph of the raw data? Typically, a plot of data from this kind of experiment is shaped as shown in Figure 3.1. Interpretation of data plotted in this form is difficult. However, a straight line is a manageable entity for biologists, statisticians, and mavens because the relationship between the independent variable and the dependent variable is easy to see or calculate.

How can raw data be made linear? Transformation of the units of both the x and y axes is frequently necessary for binary pesticide bioassay data to fit a linear model. When only units on the x axis are transformed, the plot of the data typically becomes more symmetrical (Figure 3.2). Units on the y axis can be transformed into probits, normal equivalent deviates (NEDs),[7] logits, or some other units defined by a particular distribution function. Units on the x axis are usually converted to logarithmic values; however, this transformation is not always necessary. For example, units of temperature or radiation tested in commodity treatments often need not be transformed at all for the data to fit the linear model.

In a quantal response bioassay, the points along the regression line are such that, for a given dose or concentration on the x axis, there is a corresponding probability level of response. The dose or concentration that corresponds with a specific probability of response level is the lethal dose (LD) or lethal concentration (LC). For example, LD_{50} (or LC_{50}) is the LD (or concentration) that is expected to cause 50% mortality. Depending on the definition of *dose*, other terms in the scientific literature are LC, ED (effective dose), ID (inhibitory dose or dose necessary for $x\%$ inhibition), and LT (lethal time necessary for $x\%$ mortality). The way a particular bioassay should be done (i.e., the experimental design) to estimate a given LD or LC depends on the response level of interest. Because Dr. Maven is using a statistical model to obtain this dose, *estimation* is the correct terminology; *measurement* or

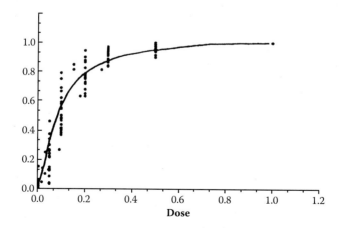

Figure 3.1 Typical plot of raw data from a dose–response experiment.

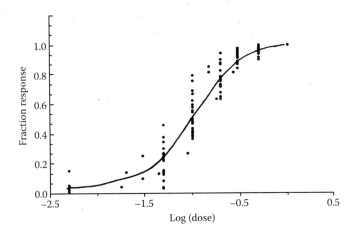

Figure 3.2   S-shaped curve that typically results from transformation of dose to logarithm.

determination should not be used in this context because she neither measures nor determines LD.

3.2 STATISTICAL METHODS

3.2.1 Probit or Logit Regression

Of the numerous types available,[8] the two distribution functions most commonly used in Equation 3.1 are normal (Figure 3.1) and the logistic (Figure 3.2). The normal distribution X is assumed in probit analysis, so that the model specified by Equation 3.1 is $p_i = \Phi(\alpha + \beta x_i)$, where Φ is the standard normal (Gaussian) distribution function. The "linear predictor" $\alpha + \beta x$ is called the probit or logit line depending on whether the normal or the logistic model is used. The parameters α and β are the intercept and the slope of the probit or logit line.

For the logit model:

$$P_i = \frac{1}{1 + e^{-(\alpha + \beta x_i)}} = \frac{e^{\alpha + \beta x_i}}{1 + e^{\alpha + \beta x_i}}$$

where e is the base of the natural logarithm ($=2.71828$).

The NED model also assumes the normal distribution, but the units on the y axis are transformed into probit +5. Traditionally, the NED model was used to simplify calculations. Because most calculations are now done with computers, this simplification is no longer necessary. Nevertheless, some computer programs such as SAS PROBIT[9] still use NED units. Therefore, estimates of the intercept obtained with the

Table 3.1 Comparison of Results of Probit (P) and Logit (L) Analyses of Dose–Response Data for Mexacarbate Applied to Western Spruce Budworm

Generation	N	LD$_{50}$		LD$_{90}$	
		P	L	P	L
F$_{11}$	290	0.83	0.83	3.32	3.21
F$_{13}$	494	0.72	0.73	2.95	2.90
F$_{11}$	180	0.86	0.89	2.46	2.49
F$_{19}$	484	0.95	0.94	3.14	3.15
F$_{20}$	364	0.64	0.64	1.84	1.79

Note: Part of these results were reported in Savin, NE, Robertson JL, Russell RM, *Bull. Entomol. Soc. Am.* 23, 257–266, 1977.

SAS probit program are always 5 units smaller than those obtained with a program (such as POLOSuite[10]) that uses probits.

By convention based solely on tradition, entomologists have tended to use the probit model. Although a debate about use of one model versus the other has continued among statisticians,[5,11,12] no scientific biological evidence has ever been presented to show that bioassay data are always distributed as either a normal or a logistic function.

Does the use of one model or the other matter? This question is difficult or impossible to answer because similar results are obtained with either one except at the extreme ends of the probability distribution. For example, Savin et al.[13] have shown that, for mexacarbate applied to laboratory-reared (nondiapausing) western spruce budworm, LD$_{50}$ estimates of LD$_{50}$s for probit and logit models varied by 1.4% and LD differed LD$_{90}$s varied by 0.48%. At the LD$_{99}$, however, logit values were consistently higher (25%). Results from probit and logit analysis of the same data sets are shown in Table 3.1. Clearly, estimates at 50% and 90% levels of mortality are very close, regardless of the model used.

The parameters of each model can be estimated by maximizing a loss function. Two popular loss functions are (1) minus the logarithm of the likelihood function (maximum likelihood [ML] procedure) and (2) the χ^2 function (minimum X^2 method). A third function is the deviance,[13] which is a standardized log-likelihood function.

3.2.1.1 Goodness of Fit

The measure of how well data fit the assumptions of the model (whether probit, logit, or some other related model; whether error is binomial) is called *goodness of fit*. The usual way to test fit is with a χ^2 test. In this test, values (responses) predicted by the model are compared with values actually observed in the bioassay. If the values differ significantly at $P = 0.05$, the model is assumed not to fit the data and a more appropriate model should be sought. However, alternatives to the probit or logit models are not easy to use. (The need for further research and development in this aspect of pesticide bioassay techniques is further discussed later in this chapter.)

Some scientists abandon the probit or logit model and examine response at each dose level separately. Others (Williams[14] and SAS[9]) multiply the variances by a heterogeneity factor ($\chi^2/(k-2)$, where k is the number of doses) to account for extra variation that causes poor fit. However, neither of these alternatives provides any clue about why the data do not fit the model. A more meaningful approach is to examine possible causes of lack of fit by means of residual plots (Preisler[15]).

A *residual* is the difference between the observed value and the expected value. Because binomial responses do not have a constant variance over the range of responses tested, they must be standardized by division by their standard errors, $\sqrt{np(1-p)}$. The variable n is the number of subjects tested at the specific dose and p is probability. For good fit, residuals plotted against predicted values usually lie within a horizontal band around zero (mostly between –2 and 2).

Preisler[15] identified five causes of significant departure from the probit binomial model that produce large χ^2 values. First, outliers that do not fit the model might be the cause of lack of fit, especially if they occur at a response level near 0% or 100%. An outlier should be discarded from the data set only if strong evidence suggests that a recording error has occurred. Otherwise, the outlier may indicate interaction with or the existence of an important independent variable. The second cause may be omission of a significant explanatory variable from the model (for example, when a multiple probit or logit model is appropriate but a univariate model is used). Third, the statistical model might not fit the data. For example, a probit regression line with a plateau is characteristic of genetically heterogeneous populations and is often observed in groups that contain resistant and susceptible individuals.[16]

Responses related to some factor in the model may cause correlation between test subjects. As a result, error terms are not independent. Thus, insects from the same generation might be more alike in their response to a pesticide than they are compared with insects in another generation. A large data set for experiments in which three pesticides—mexacarbate, pyrethrins, and dichlorodiphenyltrich loroethane—were tested on more than 40 generations of a laboratory colony of nondiapausing western spruce budworm[17] shows this type of lack of fit. In Figure 3.3, clustering of

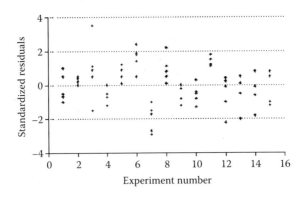

Figure 3.3 Data in which residuals are clustered.

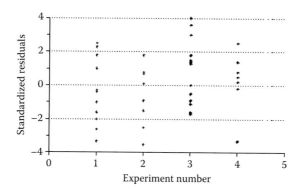

Figure 3.4 Data in which the absolute value of more than 5% of standardized residuals are >2.

residuals within each experiment is apparent. Therefore, a simple binomial model is not appropriate because use of such a model assumes that observations are independently distributed.

Finally, Preisler[15] has identified a situation in which error terms are not binomial. The data set[18] that she used was collected in tests with an acaricide, propargite, on spider mites. Even when residuals were standardized, many of their absolute values were still >2 (see Figure 3.4). The extrabinomial variability in these data could not be explained by an experiment or generation effect. Likewise, sampling from a genetically heterogeneous population did not appear to be responsible.

For lack of fit caused by lack of independence of error terms (as in the data set for western spruce budworm) and extrabinomial variability (as in the data set for spider mites), Preisler[15] has described a method to obtain ML estimates for parameters in a compound binomial probit model so that hypotheses of parallelism and equality can still be tested. This method is based on addition of a random effect factor to the probit line at the dose (or concentration) level. Inclusion of this procedure in available probit or logit statistical packages would improve the reliability of standard errors for LD estimates and the reliability of the results of tests of parallelism and equality.

3.2.1.2 Lethal Dose Ratios

Ratios at a specific response level such as LD_{50} or LD_{90} are often used to determine the relative toxicity of a number of chemicals compared with a standard chemical[19] to determine relative susceptibility of populations[20] and to study the extent,[21,22] and genetic bases,[23] for pesticide resistance. A LD ratio is more general than the alternative statistic, relative potency,[5] which assumes that the regression lines being compared are parallel. If there were some rationale by which parallelism could always be assumed to occur, relative potency could be used rather than a LD ratio. However, regression lines are rarely parallel, particularly in bioassays of several pesticides with different modes of action or among populations in the process of becoming resistant to a pesticide.

Given the fact that a simple ratio of one LD to another does not provide any estimate of the error involved in the calculation, the most practical and least restrictive alternative is to estimate 95% confidence intervals for each ratio. Based on estimates for the intercepts (α_i) and the slopes (β_i, $i = 1, 2$) of two probit (or logit) lines and estimates of their variance–covariance matrices, the 95% confidence interval for the ratio is calculated by the following steps:

1. Calculate

$$\hat{\theta}_i = \frac{x_\pi - \hat{\alpha}_i}{\hat{\beta}_i}$$

for $i = 1, 2$

where x_π is the π-th percentile point of the probit (or logit) distribution curve. Values of x_π for levels of response that are most often of interest are as follows:

π	10	25	50	75	80	90	95	97.5	99
x_π	−128	−0.67	0	0.67	0.84	1.28	1.65	1.96	2.33

For example, if $\pi = 90$, then $x_{90} = 1.28$.

2. Calculate

$$\text{vâr}(\hat{\theta}_i) = \frac{1}{\hat{\beta}_i^2}\left[\text{vâr}(\hat{\alpha}_i) + 2\hat{\theta}_i \text{côv}(\hat{\alpha}_i, \hat{\beta}_i) + \hat{\theta}_i^2 \text{vâr}(\hat{\beta}_i)\right]$$

for $i = 1, 2$.

3. Calculate

$$a = \hat{\theta}_1 - \hat{\theta}_2$$

and

$$\sigma = \sqrt{\text{vâr}(\hat{\theta}_1) + \text{vâr}(\hat{\theta}_2)}$$

4. Estimates of the ratio of two LDs and the approximate 95% confidence boundaries are given by:

$$\text{ratio} = 10^a$$

$$\text{lower boundary} = 10^{a-2\sigma}$$

$$\text{upper boundary} = 10^{a+2\sigma}$$

Relative potency between two toxicants is always the same regardless of dose level, while ratios of LDs can be very different at, for instance, LD_{50} and LD_{90}, because two lines are not parallel. Especially when dose–response regressions are used as evidence for resistance to pesticides, reporting of ratios at two response levels such as 50% and 90% is probably most informative.

3.2.1.3 Comparison of Lethal Dose Ratios

Suppose an investigator is interested in whether the LD of one group differs significantly from the LD of another group. Dr. Maven, for example, might like to know whether the LD for the parent generation of *P. giganticus* differs significantly from the LD for the second laboratory generation.

Many investigators have used a crude method to address this question. They compare LDs by examining their 95% confidence limits. If the limits overlap, then the LDs do not differ significantly except under unusual circumstances.

For Dr. Maven's results, the LD_{90} for the parent generation is 0.147, with 95% confidence limits of 0.111 to 0.253. Values for the second laboratory generation are 0.476, with limits of 0.355 to 0.818. The 95% confidence limits of these LD_{90} do not overlap, and Dr. Maven concludes that they probably differ significantly. However, the exact significance level for this procedure is not clear: it is not 5%.

Another method, and one for which significance can be set at $P = 0.05$ or any other probability level of choice, is as follows: Calculate the ratio of the LD, then calculate the 95% confidence limits for this ratio as described earlier in this chapter. If the 95% confidence interval includes 1, then the LD are not significantly different.

3.3 RISK OF ERRONEOUS CONCLUSIONS

Dr. Maven now understands several statistical tests that can be performed on the same data sets. For example, she can test whether two or more lines are the same, whether they are parallel but not equal, whether LD of interest are the same, and so on. Each test has a 5% error rate, so there is a 1-out-of-20 chance that the results of the test are wrong. Across all of the statistical tests, the chance of making a wrong statement about the parameters is also 1 out of 20 as long as the statistical tests are independent. Dr. Maven can accept this risk.

For investigators not willing to live so dangerously, the solution is simple: Use a lower error level such as 1%. However, the error level cannot simply be reduced indefinitely because, as the error level decreases, the power (or ability) of a test procedure to detect any differences between dose–response lines will approach zero.

3.3.1 Interpreting the Results of Hypotheses Tests

Despite protests to the contrary, hypothesis *tests* are as essential for the interpretation aspect of the analyses of bioassay data as is statistical hypotheses testing. As Dr. Tarleton so wisely stressed during his course, a difference in results is not a difference unless it is significant. Identification of significant differences is the interface between biology and statistics. The specific nature of hypotheses can be tested with binary quantal response bioassays. The three possible outcomes of hypothesis tests of data depend on whether Dr. Maven is interested in regression lines, point estimates (i.e., LDs) on different lines, or groups of regression lines.

3.3.1.1 Regression Lines

Hypothesis tests can be done to compare the slopes and intercepts of different probit or logit lines. Three hypotheses are usually of interest: (1) the lines are parallel but not equal, (2) the lines are equal, or (3) the lines are neither parallel nor equal (Figure 3.5a–c).

When two (or more) lines are parallel but not equal, their slopes are not significantly different (see Figure 3.5a) even though their intercepts differ significantly. Biologically, the slope of a probit or logit regression estimates the change in activity per unit change in dose or concentration. Pharmacological evidence described by Hardman et al.[24] and Kuperman et al.[25] suggests that the slope of a probit or logit regression also reflects the quality of enzymes involved in detoxification. Thus, parallel lines may indicate that organisms have qualitatively identical, but quantitatively different, levels of detoxification enzymes.[26] However, direct biochemical evidence of this interpretation is lacking. Pesticide bioassays, done in conjunction with biochemical assays of relevant detoxification mechanisms, may provide further information about this interpretation. A more definitive biochemical definition of slope would be particularly useful in studies of pesticide resistance and environmental toxicology.

Theoretically, the intercept of a probit or logit regression should correspond with the response threshold (i.e., the response that occurs with no treatment). Natural response is a statistical estimate of this response threshold; this value is provided by the response of controls in a bioassay. At the response threshold, at least one organism in the population responds to the treatment. If regressions are equal (a second possible outcome of hypothesis testing), they do not differ significantly in either slope or intercept (see Figure 3.5b). Equality may mean that detoxification enzymes in the organisms with equal responses are qualitatively and quantitatively the same. However, this interpretation has not yet been examined biochemically in arthropods. Finally, when lines are neither equal nor parallel (see Figure 3.5c), neither their slopes, nor, quite often, their intercepts are equal. Biochemically, this outcome may mean that detoxification enzymes differ qualitatively or that totally different enzymes occur in the organisms.

Physical processes, such as absorption through the cuticle or gut, target site sensitivity, or excretion, may also be relevant in inequality, parallelism, and equality. In summary, the biochemical bases and physiological processes in the arthropod that result in parallelism and equality would be useful areas for future toxicological and biochemical research.

Hypotheses tests of parallelism and equality of lines can be tested by likelihood ratio (LR) tests. The degrees of freedom in the LR test is the number of parameters

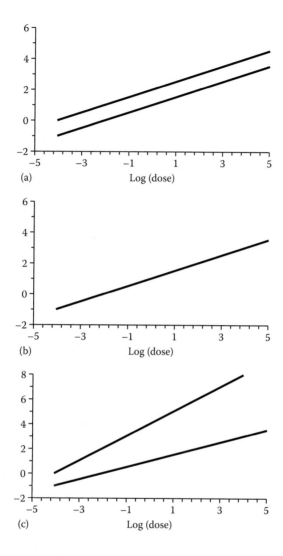

Figure 3.5 The three possible results of tests of the hypothesis of parallelism and the hypothesis of equality. (a) Two (or more) lines are parallel, but not equal, (b) lines are equal, or (c) lines are neither parallel nor equal.

in each line (=2 for slope and intercept), multiplied by the total number of individual lines, minus the number of parameters in the composite line (=2). In the LR test of parallelism, slopes of individual lines from different experiments are assumed to be the same; each line is then compared with the composite line. In this test, the degrees of freedom is the number of individual lines minus 1. As will be illustrated in Chapter 4, LR tests of equality and parallelism are done automatically by the PoloSuite software program.[27]

3.3.1.2 Point Estimates

Suppose that Jessica has completed a large study of the responses of Godfather worms to geographic isolates of a virus. Dr. Maven is interested in whether the LD_x of one isolate differs significantly from the LD_x of another and which isolates merit full-scale production. In the past, the usual way to compare LDs or other point estimates was to examine their 95% confidence limits.[28] If the limits overlapped, then the LDs were not considered to be significantly different except under unusual circumstances. Based on the evidence provided by Wheeler et al.,[29] we recommend that this test not be used because it lacks statistical power. By using Monte Carlo simulation, these authors[29] provide valuable evidence that the ratio test is far preferable to the overlap test based on statistical power of the test and better type II error rates. The ratio test, for which significance can be set at $P = 0.05$ or any other probability level of choice, is done as follows. Calculate the ratio of the LDs; then calculate the 95% confidence interval for this ratio as shown in Section 3.2. If the 95% confidence interval includes 1, then the LDs are not significantly different.

3.3.1.3 Groups of Lines with Equal Response

LR tests of equality and parallelism are perhaps the most efficient means to compare probit or logit regressions, assuming, of course, that the data fit the probit or logit model. An alternative method is use of a weighted analysis of variance. One example of the use of this method is separation of response lines of Indianmeal moth, *Plodia interpunctella* (Hübner), to the microbial pesticide *Bacillus thuringiensis* Berliner.[30] However, analysis of variance ignores information such as the fact that dose–response curves are monotonically increasing, S-shaped curves. Weighted analysis of variance is also computationally less efficient than ML estimation. Tabashnik et al.[31] have argued that analysis of variance is preferable for studies of pesticide resistance because, among other reasons, it avoids some of the problems that occur when probit analysis is used for detecting and monitoring resistance. These problems are described by Roush and Miller.[32] One of the assertions made by Tabashnik et al.[31] is that analysis of variance provides a direct, overall test of the hypothesis that populations differ in their response.

Separation of probit or logit regressions into groups has merit, especially when the results concern insecticide resistance or other areas of research in which genetic shifts in populations of a target organism or genetic differences in a microbial pesticide affect the responses estimated by probit or logit regression. Examples of useful information to be gained by grouping dose–response lines are (1) detection of significant shifts in a species' response in a particular geographic area in relation to continued selection pressure, (2) comparison of activities of isolates of a particular strain of virus or bacterium on a target organism (see, e.g., Shapiro et al.[33]), and (3) detection of geographic variation in a species' response to a pesticide. No statistical tests analogous to multiple range tests have been developed for use with probit or logit regressions. The only procedure available consists of a sequence of tests of the hypothesis of equality (e.g., Robertson and Stock[34] and Shapiro et al.[33]). Each test must be done at a level of probability lower than $P = 0.05$ so that the overall $P = 0.05$.

3.4 ALTERNATIVES TO PROBIT AND LOGIT ANALYSIS

Both probit and logit analyses are based on two-parameter, symmetric toler-ance distributions described by Finney[5] and Cox.[35] Copenhaver and Mielke[36] have described another method, quantit analysis, that is based on a three-parameter fam-ily of symmetric tolerance distributions. The Ω distribution includes both the nor-mal distribution used in probit analysis and the logistic distribution used in logit analysis. Thus, the normal and logistic distributions are subsets of the Ω distribution. Possibly because the computer program available for quantit analysis was written in FORTRAN for IBM mainframe computers and the output is fairly difficult to inter-pret, quantit analysis has not been used widely for analysis of bioassay data.

Because the Ω distribution is symmetric, LD_{50} values estimated by probit, logit, or quantit analyses do not differ appreciably. However, differences at LD_{99} are large. As shown by Copenhaver and Mielke,[36] the Ω model represents a more general-ized symmetric tolerance distribution that may give better fit in the extreme tails of response, such as below the LD_5 and above the LD_{95}. However, many of the data sets used as examples include too few test subjects to be considered well designed even with perfect dose placement. Furthermore, quantit analysis is, like probit and logit analysis, inflexible in the sense that it is based on symmetric distribution functions.

As will be described in Chapter 11, the two-parameter complimentary log–log (CLL) model[37] may offer even greater flexibility than does the Ω model because it includes asymmetric tolerance distributions but can give approximations of sym-metric distributions such as the normal, logistic, and Ω models. Future research in statistical analyses of binary quantal response data should be directed at the devel-opment of more flexible methods, such as CLL analysis, for practical use by biolo-gists. Finally, Hubert[38] describes numerous nonparametric models; these may merit further consideration for use in the analysis of data from univariate quantal response bioassays.

REFERENCES

1. Singh, P. and Moore, R. F., *Handbook of Insect Rearing, Vols. I and II*, Elsevier, Amsterdam, 1985.
2. Cohen, A. C., *Insect Diets: Science and Technology*, CRC Press, Boca Raton, FL, 2004.
3. Robertson, J. L., Russell, R. M., Preisler, H. K., and Savin, N. E., *Bioassays with Arthropods*, 2nd ed., CRC Press, Boca Raton, FL, 2007.
4. Stark, J. D. and Banks, J. E., Population-level effects of pesticides and other toxicants on arthropods, *Annu. Rev. Entomol.* 48, 505, 2003.
5. Finney, D. J., *Probit Analysis*, Cambridge University Press, Cambridge, England, 1971.

6. Hamilton, M. A., Estimation of the typical lethal dose in acute toxicity studies, in Krewski, D. and Franklin, C., *Statistics in Toxicology*, Gordon and Breach Science Publishers, New York, 1991, pp. 61–68.

7. Gaddum, J. H., Reports on biological standards. III. Methods of biological assay depending on quantal response, *Spec. Rep. Ser. Med. Res. Coun. Lond.* 183, 1933.

8. Hastings, N. A. J. and Peacock, J. B., *Statistical Distributions, A Handbook for Students and Practitioners*, Butterworths, London, 1974.

9. SAS Institute, http://www.sas.com/technologies/analytics/statistics/stat.

10. LeOra Software, *PoloSuite*, for Windows and OS, LeOra Software LLC, PO Box 562, Parma, MO 63870, 2016.

11. Berkson, J., Why I prefer logits to probits, *Biometrics* 7, 327, 1951.

12. Copenhaver, T. W. and Mielke, P. W., Quantit analysis: A quantal assay refinement, *Biometrics* 33, 175, 1977.

13. Savin, N. E., Robertson, J. L., and Russell, R. M., A critical evaluation of bioassay in insecticide research: Likelihood ratio tests of dose-mortality regression, *Bull. Entomol. Soc. Am.* 23, 257–266, 1977.

14. Williams, D. A., Extra-binomial variation in logistic linear models, *Appl. Statistics* 31, 144, 1982.

15. Preisler, H. K., Assessing insecticide bioassay data with extra-binomial variation, *J. Econ. Entomol.* 81, 759, 1988.

16. Tsukomoto, M., The log-dosage probit mortality curve in genetic researches of insect resistance to insecticides, *Botyu-Kagaku* 28, 91, 1963.

17. Savin, N. E., Robertson, J. L., and Russell, R. M., A critical evaluation of bioassay in insecticide research: Likelihood ratio tests of dose-mortality regression, *Bull. Entomol. Soc. Am.* 23, 257, 1977.

18. Grafton-Cardwell, E. E., Granett, J., and Leigh, T. F., Spider mite species (Acari: Tetranychidae) response to propargite; the basis for a resistance management program, *J. Econ. Entomol.* 80, 579, 1987.

19. Kontsedalov, S., Weintraub, P. G., Horowitz, A. R., and Isaaya, I., Effects of insecticides on immature and adult western flower thrips (Thysanoptera: Thripidae) in Israel, *J. Econ. Entomol.* 91, 1067, 1998.

20. Luttrell, R. G., Wan, L., and Knighten, K., Variation in susceptibility of Noctuid (Lepidoptera) larvae attacking cotton and soybean to purified endotoxin proteins and commercial formulations of *Bacillus thuringiensis*, *J. Econ. Entomol.* 92, 21, 1999.

21. Scharf, M. E., Meinke, L. J., Siegfried, B. D., Wright, R. J., and Chandler, L. D., Carbaryl susceptibility, diagnostic concentration determination, and synergism of U.S. populations of western corn rootworm (Coleoptera: Chrysomelidae), *J. Econ. Entomol.* 92, 33, 1999.

22. Prabhaker, N., Toscano, N. C., and Henneberry, T. J., Evaluation of insecticide rotations and mixtures as resistance management strategies for *Bemisia argentifolii* (Homoptera: Aleyrodidae), *J. Econ. Entomol.* 91, 820, 1998.

23. Bengston, M., Collins, P. J., Daglish, G. J., Hallman, V. L., Kopittke, R., and Pavic, H., Inheritance of phosphine resistance in *Tribolium casteneum* (Coleoptera: Tenebrionidae), *J. Econ. Entomol.* 92, 17, 1999.

24. Hardman, H. F., Moore, J. I., and Lum, B. K. B., A method for analyzing the effect of pH and ionization of drugs upon cardiac tissue with special reference to pentabarbitol, *J. Pharmacol. Exp. Ther.* 126, 136, 1959.

25. Kuperman, A. S., Gill, E. W., and Riker, W. F., The relationship between cholinesterase inhibition and drug-induced facilitation of mammalian neuromuscular transmission, *J. Pharmacol. Exp. Therap.* 132, 65, 1961.

26. Robertson, J. L. and Rappaport, N. G., Direct, indirect and residual toxicities of insecticide sprays to western spruce budworm, *Choristoneura occidentalis* (Lepidoptera: Tortricidae), *Can. Entomol.* 111, 1219, 1979.

27. LeOra Software, *PoloSuite*, POLO for Windows and Mac OS, LeOra Software LLC, PO Box 562, Parma, MO 63870. See http://www.LeOra-Software.com.

28. Payton, M. E., Greenstone, M. H., and Schenker, N., Overlapping confidence intervals or standard error intervals: What do they mean in terms of statistical significance? http://www/insectscience.org/3.34.

29. Wheeler, M. W., Park, R. M., and Bailey, A. J., Comparing median lethal concentration values using confidence interval overlap or ratio tests, *Environ. Toxicol. Chem.* 25, 1441, 2006.

30. McGaughey, W. H., Insect resistance to the biological insecticide *Bacillus thuringiensis*, *Science* 229, 193, 1985.

31. Tabashnik, B. E., Cushing, N. L., and Johnson, M. W., Diamondback moth (Lepidoptera: Plutellidae) resistance to insecticides in Hawaii: Intra-island variation and cross-resistance, *J. Econ. Entomol.* 80, 1091, 1987.

32. Roush, R. T. and Miller, G. L., Considerations for design of insecticide resistance monitoring programs, *J. Econ. Entomol.* 79, 293, 1986.

33. Shapiro, M., Robertson, J. L., Injac, M. G., Katagin, K., and Bell, R. A., Comparative infectivities of gypsy moth (Lepidoptera: Lymantriidae) nucleopolyhedrosis virus isolates from North America, Europe, and Asia, *J. Econ. Entomol.* 77, 153, 1984.

34. Robertson, J. L. and Stock, M. W., Toxicological and electrophoretic population characteristics of western spruce budworm, *Choristoneura occidentalis* (Lepidoptera: Tortricidae), *Can. Entomol.* 117, 57, 1985.

35. Cox, D. R., *Analysis of Binary Data*, Methuen, London, 1970.

36. Copenhaver, T. W. and Mielke, P. W., Quantit analysis: A quantal assay refinement, *Biometrics* 33, 175, 1977.

37. Preisler, H. K. and Robertson, J. L., Analysis of time-dose-mortality data, *J. Econ. Entomol.* 82, 1534, 1989.

38. Hubert, J. J., *Bioassay*, Kendall/Hunt, Dubuque, IA, 1984.

CHAPTER 4

Binary Quantal Response: Data Analyses

The calculations in probit or logit analyses involve many complex equations that are described by Finney.[1] In the early 1960s, before personal computers and programs written to do the calculations existed, analyses of bioassay data were a horrendous chore. Now, several well-documented computer programs are available to perform the data in a minimum amount of time. Most of these early programs simply resembled the manual computation scheme and were written by programmers who made no attempt to take full advantage of computer technology. Unfortunately, two of the first computer programs[2,3] used by many entomologists amplified how easily mistakes can remain undetected and how the programs can be widely used until a subsequent careful examination reveals major problems. The errors in these programs were not detected until the output was needed for types of tests that had not been done before.[4]

Among the problems in the more popular of the two early programs, 14 standard errors of the slopes were computed incorrectly and were in error as much as 50%. Because these values were incorrect, the confidence limits for the lethal doses (LDs) were also erroneous. Worst of all, results were often invalid but the printout did not say so.

Several well-documented probit programs or procedures are now available. Among these are SAS,[5] GENSTAT,[6] BLISS,[7] GLIM,[8] and POLO.[4,9–11] All of these programs were originally written for large, mainframe computers; SAS and POLO[11] are also available for use on personal computers. However, all of the programs, except POLO, produce output that is not easily read or interpreted by a biologist.

4.1 PoloJR

The POLO[4] software program was written to do the computations described in Finney's *Probit Analysis*,[1] but with two unique additions. First, logit analysis was an option. Second, and of most practical importance to biologists, likelihood ratio (LR) tests of equality and parallelism were included. POLO-PC[9] produced an identical output. PoloJR[11] is computationally identical to its predecessors. However, when the program was redone for Microsoft Windows and MacIntosh computers, new features were added and unnecessary details were eliminated from the output. Among the new features is the ability to add data directly into the program rather than importing from a text file. Two ways of displaying goodness-of-fit tables of standardized residuals, and their plots, were again included to identify sources of poor fit to the probit or logit model. The residual plot includes lines at the +2 and −2 y axis to provide easier recognition of outliers.

Of the commercially available probit or logit programs, PoloJR is one of the very few that can be easily used and understood by users who are not statisticians. Statistical terminology in the program is adapted from that used by Finney.[1] Unfortunately, some terms cannot be explained in words and must be described by equations, which will be described here.

4.1.1 Data

The main screen (Figure 4.1) has drop-down menus for each of four programs: PoloJR, PoloMulti, PoloMixture, and OptiDose. Instead of entering the data manually, Jessica selects "Open a data file" and all of the available files in the PoloSuite directory are listed (Figure 4.2). When "Bugchlor.txt" is selected, the data file for the experiment is listed on the screen (Figure 4.3). In this file, each data set consists of

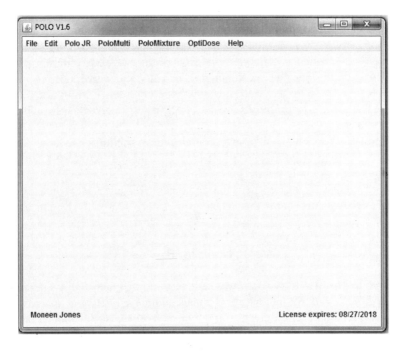

Figure 4.1 PoloSuite opening screen.

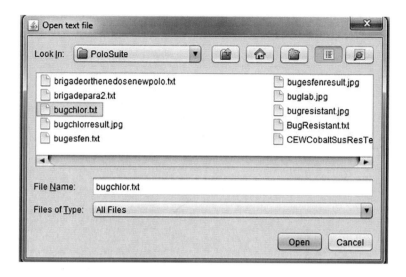

Figure 4.2   PoloSuite ready to open a data file.

Bug, chlorantraniliprole
Oct 20 2015 Lab Colony
*lab_col
0.0 203 35
0.01 200 34
0.03 200 46
0.1 199 144
0.3 201 185
1.0 203 200

Figure 4.3   PoloSuite displaying data file.

two title lines, followed by colony description (*) lines that describe the actual data. In the preparation line, only the first eight characters are shown. Lab colony is the description for the baseline population data.

Under a given title, any number of data sets can be analyzed. In each data line, only one line per dose can be used. Each data line contains three fields in specific order. The first is dose (x), the second is the number (n) of test subjects, and the third is the number that respond (r). A sample display of data is shown in Figure 4.4. The results of each bioassay can be summarized manually but should be entered separately. Entry of the separate data points without manual summarization minimizes the introduction of errors in addition and provides the correct degrees of freedom for the analysis.

S.No.	Preparation	Dose (d)	# of subjects (n)	# that responded (r)
1	lab_col			
2		0.0	203.0	35.0
3		0.01	200.0	34.0
4		0.03	200.0	46.0
5		0.1	199.0	144.0
6		0.3	201.0	185.0
7		1.0	203.0	200.0

View Data
Bug, chlorantraniliprole
Oct 20 2015 Lab Colony

Figure 4.4   PoloSuite displaying data file.

4.1.2 Choice Screens and Program Options

In the next dialog box (Figure 4.5), four sets of selections can be made. The first choice is that of model (probit or logit). Jessica selects a probit model. Is natural response a parameter? She selects yes. Doses can be converted to logarithm. Next, she chooses to estimate the LD_{50} and LD_{90} for the data. From the drop-down menu, "Calculate" is selected. PoloSuite automatically checks the data for correct input. If there is an error present, a pop-up message will appear. Sample errors could be the

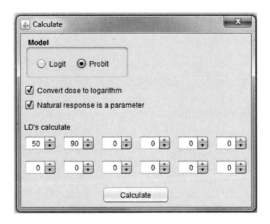

Figure 4.5   PoloSuite showing choose options window.

result of entering the number of responding individuals greater than the number of individuals tested. Everything appears to be entered correctly.

4.1.3 Display Results

The "Results" screen appears in the next window. The output that appears on the screen is shown in Figure 4.6. An option to print the data as a .txt file is available. To save output, select data using mouse, copy (ctrl C), and paste into Excel, Word, or other data program.

In this example, we chose the probit model, natural response is a parameter, doses are converted to logarithms, the number of preparations is 5, the total number of dose groups is 1, 2 LDs were estimated, and these LDs were at the 50% and 90% levels.

The first subsection for each preparation's printout gives the values, standard errors, and t-ratio for the intercept (row labeled with the preparation name), natural response if present, and slope. Caution: if the t-ratio of any slope is less than 1.96, the regression is not significant and the treatment has no effect, and the data set must be excluded from further analysis.

The χ^2 goodness-of-fit subsection lists the values of the data with each values expected value, residual, probability, and standard errors. The residuals (last column) can be plotted for further examination of lack-of-fit.

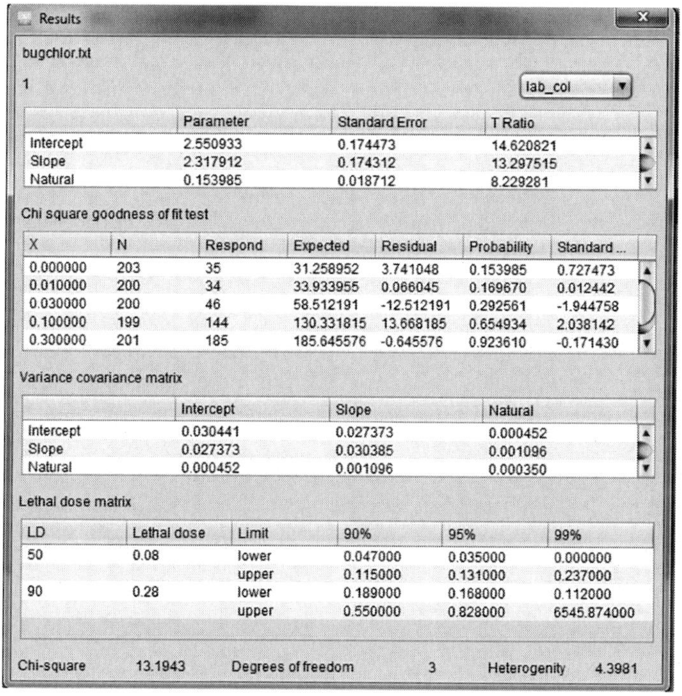

Results

bugchlor.txt

1   lab_col ▼

	Parameter	Standard Error	T Ratio
Intercept	2.550933	0.174473	14.620821
Slope	2.317912	0.174312	13.297515
Natural	0.153985	0.018712	8.229281

Chi square goodness of fit test

X	N	Respond	Expected	Residual	Probability	Standard ...
0.000000	203	35	31.258952	3.741048	0.153985	0.727473
0.010000	200	34	33.933955	0.066045	0.169670	0.012442
0.030000	200	46	58.512191	-12.512191	0.292561	-1.944758
0.100000	199	144	130.331815	13.668185	0.654934	2.038142
0.300000	201	185	185.645576	-0.645576	0.923610	-0.171430

Variance covariance matrix

	Intercept	Slope	Natural
Intercept	0.030441	0.027373	0.000452
Slope	0.027373	0.030385	0.001096
Natural	0.000452	0.001096	0.000350

Lethal dose matrix

LD	Lethal dose	Limit	90%	95%	99%
50	0.08	lower	0.047000	0.035000	0.000000
		upper	0.115000	0.131000	0.237000
90	0.28	lower	0.189000	0.168000	0.112000
		upper	0.550000	0.828000	6545.874000

| Chi-square | 13.1943 | Degrees of freedom | 3 | Heterogenity | 4.3981 |

Figure 4.6 PoloSuite showing choose options window.

The variance–covariance matrix follows. These numbers are used to estimate the 95% confidence intervals for ratios and identify significant differences between the first and subsequent preparations.

The LD matrix subsection shows the lower and upper limits for 90%, 95%, and 99% confidence levels. If the statistic g (Finney 1971) is greater than 0.5 at 0.90, 0.95, or 0.99, that confidence limit will not be printed.

Values of χ^2, degrees of freedom, and heterogeneity factor (which equals the χ^2 divided by degrees of freedom) follow. When the heterogeneity factor is greater than 1.0, a plot of the data should be examined because the data do not fit the model. Such a plot may reveal systematic departure from linear regression, in which case a function other than logarithm of dose may be more appropriate.

A plot of the dose–response curve for the Lab Colony (Figure 4.7) is created by Jessica. First, probits or logits of response must be plotted. If the shape of the plot is linear with dose (or concentration) on the x axis, no transformation of dose should be used. If the plot is not linear, a meaningful transformation should be chosen so that the data points seem to lie on a line. She also examines a plot of the residuals (Figure 4.8) to identify any outliers or other possible causes of lack-of-fit. In this case, the outliers are within −2 and +2 boundaries, so all looks good.

Now, Jessica wants to examine differences in tolerance between two populations of the Godfather larvae—a population thought to be resistant to pyrethroids collected 20 miles east of Schaefferville and the susceptible laboratory population

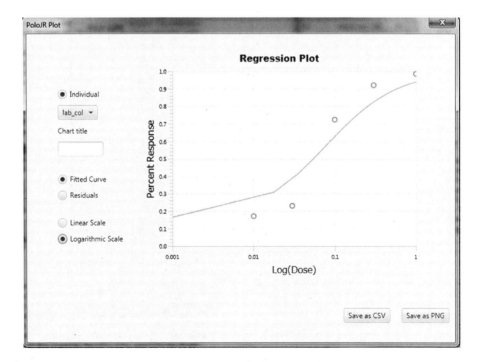

Figure 4.7 Dose–response curve for lab colony.

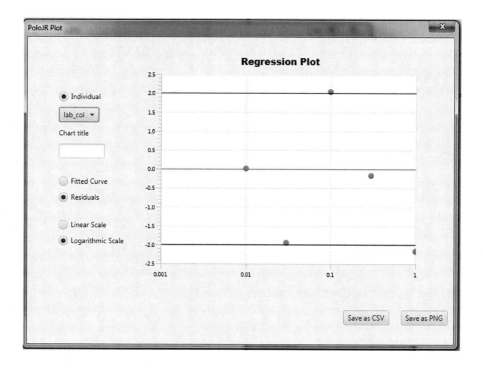

Figure 4.8   Plot of the residuals for lab colony.

that has had no contact with insecticides. The results from the experiment with the header designation "bugresistant.txt" are shown next with responses from the two populations, BugRes and BugLab to a pyrethroid (Figures 4.9 and 4.10). Because she wishes to measure the differences between populations with regard to their tolerance of this chemical, the lab population (i.e., susceptible) data are listed below the data for the resistant population.

```
BugResis, Pyrethroid
Jun 30 2015
*BugRes
0.0  60  3
3.0  60  9
10.0  60  19
20.0  60  32
40.0  60  38
50.0  60  46
*BugLab
0.0  60  5
0.03  30  7
0.1  30  7
0.3  30  6
1.0  30  3
3.0  30  3
7.0  30  10
10.0  60  32
15.0  30  22
20.0  30  30
```

Figure 4.9   Data for lab population (BugLab) versus suspected tolerant (BugRes) population.

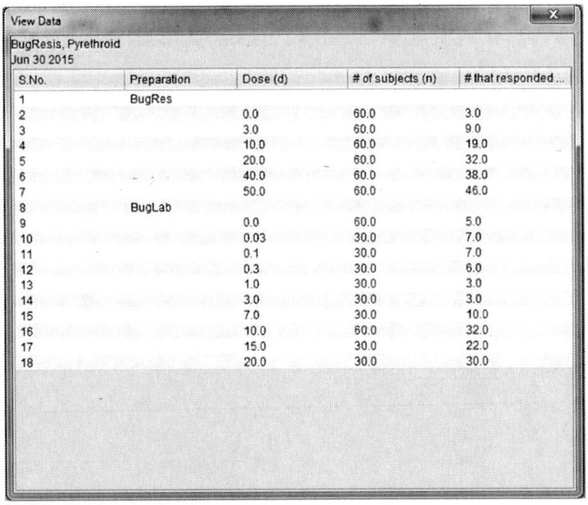

View Data				X
BugResis, Pyrethroid Jun 30 2015				
S.No.	Preparation	Dose (d)	# of subjects (n)	# that responded...
1	BugRes			
2		0.0	60.0	3.0
3		3.0	60.0	9.0
4		10.0	60.0	19.0
5		20.0	60.0	32.0
6		40.0	60.0	38.0
7		50.0	60.0	46.0
8	BugLab			
9		0.0	60.0	5.0
10		0.03	30.0	7.0
11		0.1	30.0	7.0
12		0.3	30.0	6.0
13		1.0	30.0	3.0
14		3.0	30.0	3.0
15		7.0	30.0	10.0
16		10.0	60.0	32.0
17		15.0	30.0	22.0
18		20.0	30.0	30.0

Figure 4.10   Data view for BugResistant.txt.

4.1.3.1 Parameters

The results are shown in Figure 4.11a and b. The first section lists the parameters for the analysis of the data file. In this example, Jessica chose the probit model, natural response is a parameter, doses are converted to logarithms, the number of preparations is 5, the total number of dose groups is 2, and 2 LDs were estimated at the 50% and 90% levels.

4.1.3.2 Individual Regressions

For each of these regressions, intercepts and slopes are always unconstrained. Intercepts and slopes "unconstrained" means that the data for each preparation are analyzed separately with the probit model using maximum likelihood (ML) procedures. ML is an estimation procedure based on the joint probability of all the observations. A later investigation by Smith et al. suggested that use of the ML method was preferable to the minimum X^2 procedure developed by Berkson[12] and this method was used in POLO, PoloPlus, and PoloJR for simplicity. Constrained slopes and intercepts are used only for the hypotheses tests to be described in Section 4.1.3.3.

The next subsection for each preparation's output gives the values, standard errors, and t-ratio for the intercept (row labeled with the preparation name), natural response (if present), and slopes. The t-ratio equals the parameter estimate divided by its standard error. The value of the t-ratio of the slope is crucial: if the t-ratio of any slope is <1.96 (5% significance point for the t *distribution with* ∞ *df*),

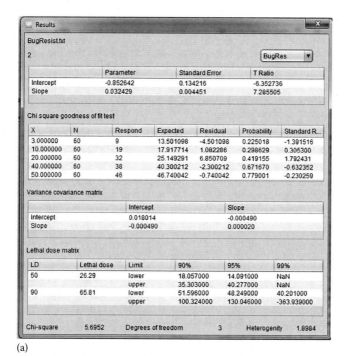

Figure 4.11 Results for (a) BugRes and (b) BugLab populations.

the regression is not significant and the treatment has no effect, and the data set must be excluded from further analysis.

The t-ratio test is not the same as the goodness-of-fit test (described in the following), which tests how well the data fit the probit or logit model. In simplest terms, the t-ratio test on the slope parameter confirms (or denies) the existence of a dose–response line, whereas goodness-of-fit tests how well the data fit the model assumed in the analysis.

The variance–covariance matrix numbers are used to estimate the 95% confidence intervals for ratios and to identify significant differences between the first and subsequent preparations.

The χ^2 goodness-of-fit test subsection shows how well each data set fits the probit model. Goodness of fit must be evaluated. If data do not fit the probit or logit model, residuals should be calculated and graphed. If overdispersion is detected, use of a compound binomial model based on the addition of a random effect factor to the probit or logit line at the level of the dose or concentration should be considered.

A residual is the difference between the observed value and the expected value (for example, $X = 3.0$ with n of 30 and r of 3 has an *expected* mortality of 7.06 with a difference or residual of -4.06). Residuals are used to identify causes of lack of fit. If the model fits, the value of χ^2 (Figure 4.11a and b) will be less than the tabular value for the appropriate degrees of freedom. The tabular value ($P = 0.05$) of χ^2 for 3, 7 degrees of freedom is 14.07, 7.8, respectively; thus, both of Dr. Maven's data sets (for the lab and field generations) fit the assumptions of the probit model adequately.

The standardized residuals (last column) should be plotted for further examination of lack of fit. Because binomial responses do not have a constant variance over the range of responses tested, they must be standardized by division by their standard errors, $\sqrt{np(1-p)}$. The variable n is the number of subjects tested at the specific dose and p is probability. For a good fit, standardized residuals plotted against predicted values usually lie within a horizontal band around zero (with approximately 95% of them between -2 and 2).

Values of χ^2, degrees of freedom, and the *heterogeneity factor* (which equals the χ^2 divided by degrees of freedom) are the last calculations. PoloJR uses the heterogeneity factor as a correction when the value of χ^2 is greater than the appropriate tabular value. Formulas involved are given by Finney.[1] When the heterogeneity factor is >1.0, a plot of the data should be examined because the data do not fit the model. Such a plot may reveal systematic departure from linear regression, in which case a function other than logarithm of dose may be more appropriate. Plots of the residuals should also be examined to identify an outlier or other possible causes of lack of fit. Both types of plots can be done with PoloJR.

Estimated LDs with upper and lower confidence limits at 90% and 95% (and 99% if they can be calculated) are listed in the last subsection of the output for individual preparations. In these examples, the second column ("lethal dose") gives estimates of the LD_{50} and LD_{90}. The upper and lower limits at 0.90, 0.95, and 0.99 are the upper and lower 90%, 95%, and 99% confidence limits for the two LDs. If the statistic g (Finney[1]) is >0.5 at 0.90, 0.95, or 0.99, that confidence limit will not be printed.

4.1.3.3 Hypotheses Tests

The LR test of equality (Figure 4.8) tests whether slopes and intercepts of the regression lines are the same. If so, the lines (i.e., treatment or population effects) are not significantly different. If the hypothesis is rejected, the lines are significantly different as in this example. The statistics used in the test are the LR, degrees of freedom, and the tail probability.

The LR test of parallelism (Figure 4.8) tests whether slopes of the lines are the same; results are given in the same format as the test of equality. In this test, slopes of each line are constrained to be the same. Each line is then compared with a composite line. For these two populations, the hypothesis of parallelism is rejected and the lines have slopes that are significantly different. Jessica and Paula conclude that the relative response rates of their populations appear to be different. If these were different chemicals in the example rather than populations, the relative potency of the chemicals would be considered to be different as well.

4.1.3.4 LD Ratios

This final section of the analysis (Figure 4.12) shows the LD ratios (relative toxicity ratios) with their upper and lower 95% confidence limits calculated in

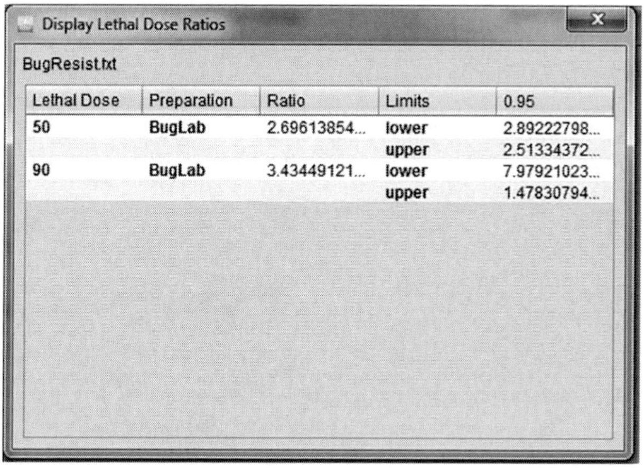

Lethal Dose	Preparation	Ratio	Limits	0.95
50	BugLab	2.69613854...	lower	2.89222798...
			upper	2.51334372...
90	BugLab	3.43449121...	lower	7.97921023...
			upper	1.47830794...

Figure 4.12 PoloJR Display of LD ratios.

comparison with the first population. These calculations were described in Chapter 3 (Section 3.2). Ratios are calculated for the same LDs estimated for the individual preparations. In this example, the LD_{50} and LD_{90} for the field population are significantly different from the LD_{50} and LD_{90} for the lab population.

4.1.4 Plot

After the calculations are finished, Jessica selects "plot" from the drop-down menu. She wants to examine the fitted curves and residuals. Each fitted curve or a plot of the residuals, plotted on a linear or logarithmic scale, can be obtained by clicking in the appropriate places. Jessica uses "Save as" menu to save the plot as a graphics file in .png format or to a csv file.

When "Next" is clicked repeatedly (with "Fitted Curve" marked), plots of the individual curves (see Figures 4.13 and 4.14) from the data are used to estimate the lines for the test of parallelism and equality (see Figure 4.15). Plots of the parallel (Figure 4.16) and collinear (Figure 4.17) cases are also shown. When "Residuals" is clicked on, the standardized residuals for individual data sets are shown (see Figures 4.18 and 4.19), followed by the standardized residuals for the tests of parallelism (see Figure 4.20) and equality (see Figure 4.21).

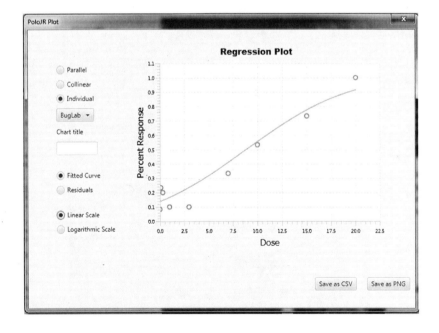

Figure 4.13   PoloJR plot of BugLab regression line.

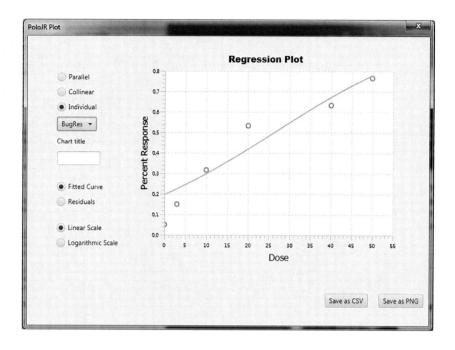

Figure 4.14   PoloJR plot of BugRes regression line.

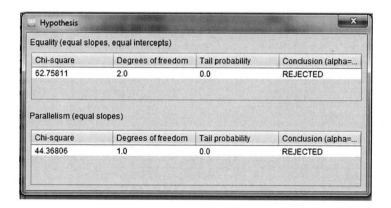

Figure 4.15   PoloPlus display of summary.

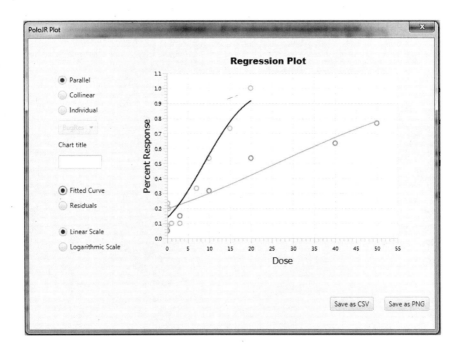

Figure 4.16   PoloJR plot of parallel case regression line.

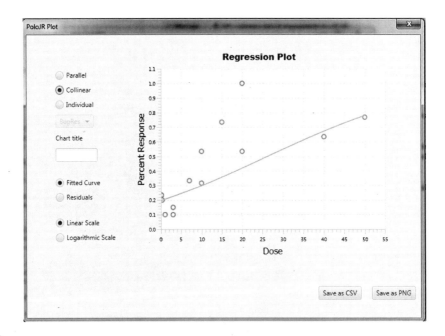

Figure 4.17   PoloJR plot of collinear case regression line.

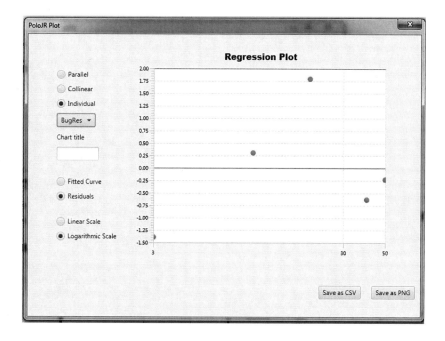

Figure 4.18 PoloJR plot of collinear case residuals.

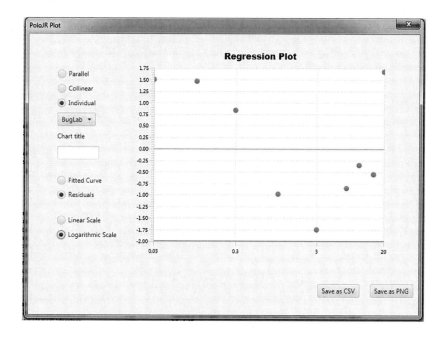

Figure 4.19 PoloJR plot of collinear case regression line.

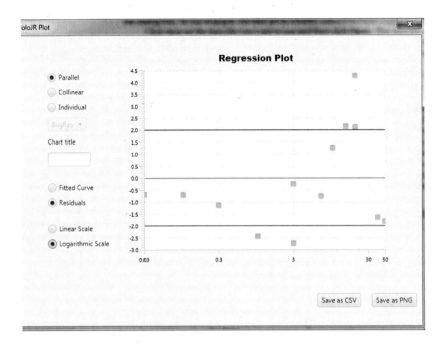

Figure 4.20   PoloJR plot of parallelism case of residuals.

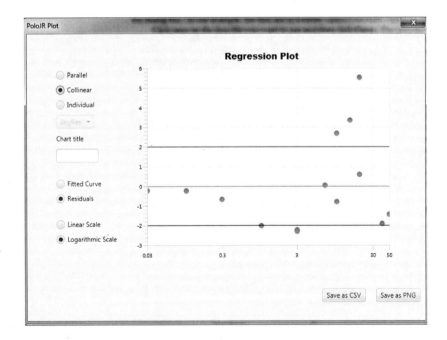

Figure 4.21   PoloJR plot of collinear case of residuals.

4.1.5 Conclusions

PoloJR produces essential information needed to analyze binary response data with accuracy and ease of interpretation. In the Windows format, the output of the original POLO, POLO-PC, and PoloPlus programs has been streamlined to include data entry and availability to MacIntosh users, thus eliminating the need to use a separate computer for calculations or another program for data entry. The carefully condensed results eliminate the intermediate calculations that tend to clutter other programs and confuse users.

4.2 SAS

The output for the probit procedure in SAS[5] exemplifies a computational scheme for rectangular data sets (defined by Fox[13] as those were originally punched on cards). Listings of results are long, detailed, and unwieldy and will not be shown here. The first part of the analysis consists of computation of the mean percent mortality at each dose. Next, a logistic estimation of the mean (μ) and the standard deviation (σ) for the normal distribution is done for each preparation. Goodness of fit by a χ^2 test is followed by a listing of first $\log_{10}$(dose) and dose with their confidence limits. Parameter estimates for calculations of toxicity ratios and their standard errors are not presented[6]; test statistics for hypotheses tests of parallelism or equality of regressions are not clearly labeled. Despite these problems, this program continues to be widely used despite the availability of other alternative programs.

4.3 R AND S-PLUS

Jessica and Paula have downloaded R[14] from the Internet and have rounded up very helpful references.[13,15,16] R[14] is a state-of-the-art, flexible software environment that can be used for all of the quantal response calculations; S-PLUS is a similar commercial package.[17] Writing the lines of code are beyond our intrepid team's capabilities, so Paula asks Dr. Tarleton to write the command lines for her to use.

With R, data must be entered in a slightly different format. In this example, results for the Godfather larvae bioassays (BugResist.txt) have been entered in a file called "Datach4new.csv" (see Figure 4.22). The R probit analysis, including tests of equality and parallelism, is as shown in Figure 4.22.

```
 1   #set directory of where file is located
 2   setwd('C:/Users/Don/Desktop/Entomology/Entomology/data/')
 3   #read in data
 4   DATA=read.csv('DATAch4new.csv',header=F)
 5   colnames(DATA)=c('DOSE','TOTAL','RESPONSE','LAB')
 6
 7   attach(DATA,2)
 8
 9   log.dose=log10(DOSE+.014)
10   LAB=factor(LAB)
11   r=RESPONSE/TOTAL
12
13   #Fit Seperate Lines
14   mod1=glm(r~LAB + LAB:log.dose-1,binomial(link='logit'),weights=TOTAL)
15   het=mod1$dev/mod1$df.residual
16   print(paste('HETEROGENEITY FACTOR =',round(het,4)))
17
18   #Test hypothesis of parallel lines
19   mod2=glm(r~LAB +log.dose-1,binomial(link='logit'),weights=TOTAL)
20   df1=mod1$df.residual;df2=mod2$df.residual
21
22   tstat=(mod2$dev-mod1$dev)/(het*(df2-df1))
23   testp=1-pf(tstat,df2-df1,df1)
24
25   paste("test deviance =",round(tstat,4)," df1=",df2-df1," df2=",df2,
26         " Pvalue=",round(testp,3))
27
28   #Final model summary
29   modf=glm(r~LAB+LAB:log.dose-1,binomial(link='logit'),weights=TOTAL)
30   summary(modf,dispersion=het,cor=F)
```

Figure 4.22   The R code for probit analysis, including tests of equality and parallelism

REFERENCES

1. Finney, D. J., *Probit Analysis*, Cambridge University Press, Cambridge, England, 1971.
2. Daum, R. J. and Killcreas, W., Two computer programs for probit analysis, *Bull. Entomol. Soc. Am.* 12, 1965, 1966.
3. Daum, R. J., Revision of two computer programs for probit analysis, *Bull. Entomol. Soc. Am.* 16, 10, 1970.
4. Russell, R. M., Robertson, J. L., and Savin, N. E., POLO: A new computer program for probit analysis, *Bull. Entomol. Soc. Am.* 23(3), 209–213, 1977.
5. *SAS User's Guide: Statistics*, SAS Institute, Cary, NC, 1982, Chap. 18.
6. Alvey, N. G., Banfield, C. F., Baxter, R. I., Gower, J. C., Krzanowski, W. J., Lane, P. W., Leech, P. K., Nelder, J. A., Payne, R. W., Phelps, K. M., Rogerts, C. E., Ross, G. J. S., Simpson, H. R., Todd, A. D., Wedderburn, R. M. W., and Wilkinson, G. N., *GENSTAT: A general statistical program*, Rothermstead Experimental Station, England, 1977.
7. Finney, D. J., *BLISS*, Department of Statistics, University of Edinburgh, Edinburgh, Scotland, 1971.
8. Baker, J. R. and Nelder, J. A., *The GLIM Manual: Release 3. Generalized Linear Modelling*, Numerical Algorithms Group, Oxford, England, 1978.
9. Robertson, J. L., Russell, R. M., and Savin, N. E., *POLO: A User's Guide to Probit OR Logit Analysis*, USDA Forest Service Gen. Tech. Rep., PSW-38, 1980.
10. LeOra Software, *Polo-Plus*, POLO for Windows, LeOra Software, 1007 B St., Petaluma, CA 94952.

11. LeOra Software, *PoloJR, POLOSuite* for Windows and Macintosh OS, LeOra Software LLC, PO Box 562, Parma, MO 63870. http://www.LeOra-Software.com.
12. Berkson, J., Why I prefer logits to probits, *Biometrics*, 7, 327, 1951.
13. Fox, J., *An R and S-Plus Companion to Applied Regression*, Sage Publications, Thousand Oaks, CA, 2002, Preface.
14. R Core Development Team, *R Reference Manual Base Package Vols. I–II*, Network Theory LTD., Bristol, England, 2004.
15. Dalgaard, P., *Introductory Statistics with R*, Springer-Verlag Inc., New York, 2002.
16. Venables, W. N. and Smith, D. M., R Core Development Team, *An Introduction to R*, Network Theory LTD., Bristol, England, 2004.
17. Solutionmetrics. What's new in S+ 8.2/8.1. http://www.solutionmetrics.com.au/products /splus/s_features.html.

CHAPTER 5

Binary Quantal Response: Dose Number, Dose Selection, and Sample Size

A year has passed and Dr. Maven, a research entomologist studying the impact of insecticides on Godfather (*Patronius giganticus*) larvae, has done so many binary response bioassays that she has actually started to trust her own expertise. As she scans her results, several things bother her. Has she used the best possible experimental design for estimating the LD_{90}s? That level is the one in which she is most interested. Has the variable sample size that she used (range, 60 to 500) affected the probit regressions? What about goodness of fit? Every once in a while, her results show vast differences among replications. And what about relative toxicities of the chemicals? Does she simply calculate a ratio of one LD to another? Or is there a way to compute the confidence limits of each ratio? These questions are the subject of this chapter.

The Maven lab is buzzing with activity. They have started several types of bioassays, among them continuous monitoring of the responses of their laboratory population; comparisons among more field populations when they can be collected; tests with chemicals, microbials; and, perhaps, physical factors; and periodic comparisons of the laboratory population, with its source populations collected from the field. In the future, surveys to determine patterns of resistance might even be necessary. Dr. Maven and her research associate, Jessica, agree that they were just plain lucky in their first tests with the laboratory and Schaefferville populations. In one of the experiments, Jessica tested nine doses with the lab susceptible population but had enough time to test only five doses with the Schaefferville population thought to be resistant (i.e., BugRes). The overall sample size for the laboratory experiments for each concentration tested was almost twice that of the other bioassay. The number of insects in the treatment groups was simply not consistent. Jessica needs a guide for dose selection and sample size because without it she could waste time, effort, and valuable insects and still not achieve the results that seem deceptively simple to obtain.

Paula begins with an extensive literature review. Although she finds that both dose placement and sample size affect the precision of estimation,[1] she quickly discovers that practical and complete guidelines have never been established for use in what she considers to be the two primary types of binary response bioassays.

The two types can be identified by both their purpose and the lethal doses (LDs) of interest. In *basic binary bioassays*, comparisons are made at LD_{50} because this is the most stable point for comparison,[2] and it is more easily estimated. LD_{90} may be included in the presentation of results to show the effect of slope on response (e.g., Robertson[3]). Examples of basic binary bioassays are those made with the same preparations tested over time or generations (e.g., Robertson et al.[4]), one or more preparations tested among populations (e.g., Robertson et al.[5]), responses of the same species among preparations (Robertson and Rappaport[6]), and comparative responses among phylogenetically related groups (e.g., Robertson et al.[7]). In *specialized binary bioassays*, more extreme LDs, such as LD_{95} or even LD_{99}, are estimated for comparative purposes. An experiment to estimate a discriminating dose that will distinguish between resistant and susceptible genotypes (e.g., Bergh et al.[8]) is such a specialized bioassay. The most extreme type of a specialized bioassay is one to estimate a probit or logit 9 ($LD_{99.9968}$); problems associated with these bioassays will be discussed in Chapter 7.

5.1 EXPERIMENTAL DESIGN

5.1.1 Dose Selection and Sample Size

Two aspects of experimental design that are paramount to the success of a binary response bioassay with one explanatory variable are dose selection and sample size: These two factors affect the precision of estimation. The entomological literature contains numerous reports (e.g., Robertson et al.[9] and Dowd and Sparks[10]) in which ratios of LDs or concentrations estimated for several chemicals at the same response level are used to compare toxicities, or toxicities of the same pesticide are compared for different life stages. In studies of pesticide resistance, a resistance ratio is often used to compare tolerances of different populations; the resistance ratio is calculated as the LD of one population divided by the LD of the most susceptible population.[11-13] Regardless of the LD, these practices assume that the value for each group has been estimated with equal precision.

Many investigators routinely use the guidelines for dose–response bioassays described by Finney,[2] without their realizing that these recommendations for dose selection pertain to the LD_{50} and not necessarily to other LDs. Likewise, methods described by Brown[14] and Freeman[15] also are applicable to the 50% response level. Tsutakawa[16] presented an approach to the problem of dose selection for efficient estimation of an arbitrary LD, but his method was not general and its computational difficulty prevented routine use. Thus, guidelines for numbers of subjects to test at each dose and for the experiment as a whole were needed.

Empirical evidence[17] suggested that different designs were required for precise estimation of LD_{50} and LD_{90}. In a limited investigation of the problems of dose number, selection, and sample size, Smith and Robertson[18] used a DOS program to examine the precision of estimates from several designs by using the criterion of width of 95% limits around LD_{50} or LD_{90}, assuming the logit model. They examined the effects of sample size and dose placement for three to eight doses and total sample sizes of 120 to 720.[1] This investigation[1] showed that at least 120 test subjects are required for a reliable dose–response experiment. Use of 60 to 64 test subjects (the smallest sample size that was tested) was not desirable. The 95% confidence limits were of poor precision in that there were frequent occurrences of confidence limits ranging from minus infinity to plus infinity and that are therefore equivalent to no confidence limits at all. Because infinite 95% confidence limits occur occasionally even with a sample size of 120, Robertson et al.[1] concluded that at least 240 test subjects should be used to estimate a dose–response relationship with acceptable precision. Is anything gained by testing >240 test subjects? Not really. The small increase in precision between 500 and, for example, 1000 probably does not justify the time or effort involved (Table 5.1). In situations in which no more than about 60 test subjects are available, an investigation can still be successful. For example, insects collected from host plants may not be as easy to obtain as those reared on artificial diets. As long as doses are selected very carefully, estimates of LDs and their 95% confidence limits can be obtained. However, Dr. Maven will always be very skeptical about the validity of results obtained when she (or anyone else) has tested <55 individuals.

Optimal placement of doses varies depending on the LD of interest and the number of doses tested (Figure 5.1). Generally, precisely estimated LD_{50}s are obtained when responses are evenly distributed between 25% and 75%. Precise LD_{90} estimates, however, require that one or two doses cause at most 10% mortality and that most doses cause between 75% and 95% mortality. However, the larger the selected value for LD, the greater the number of individuals will need to be tested to keep the 95% confidence limits from expanding exponentially whereby reducing its validity.

A computer program, OptiDose, was especially designed to provide special guidance in the placement of doses and sample sizes that are necessary for precise estimation of response levels other than 50% or 90%. The program is suited for interactive use. In the interactive mode, OptiDose can be used to explore the effects of sample size, number of doses, dose placement, and allocation of test subjects to the dose levels.

For example, Dr. Maven is planning to do a bioassay with pyrethrins. She has completed a preliminary experiment to select the application rates that cause from 5% to 99% mortality using the log scale. Estimates of the slope and intercept are 0.14 and 1.01, respectively. Jessica has had her field crew collect 180 insects for the laboratory bioassay and intends to test three to nine doses. She explores the following questions by using OptiDose interactively:

1. For precise estimation of the LD, where on the interval from 0% to 100% should the response probabilities lie?
2. What is the effect of varying the number of doses in the range of three to nine?
3. Should the same number of *Patronius giganticus* be tested with each dose, or will precision be improved by unequal allocation of insects to the doses?
4. How well can LD_{50} and LD_{90} be estimated with a design specifically selected to estimate the LD?

Table 5.1   Comparison of Designs for Estimation of an LD_{50} and LD_{90}

Doses	Cell	Sample	$P_0{}^a$	L^{*b} Probit	L^{*b} Logit
			LD_{50}		
5	25	125	0.25, 0.30, 0.50, 0.70, 0.75	0.515	0.859
	50	250		0.344	0.572
	100	500		0.237	0.394
	200	1000		0.166	0.275
	400	2000		0.116	0.193
6	25	150	0.25, 0.30, 0.35, 0.65, 0.70, 0.75	0.460	0.768
	50	300		0.312	0.518
	100	600		0.216	0.358
	200	1200		0.151	0.251
	400	2400		0.106	0.176
7	25	175	0.25, 0.30, 0.35, 0.55, 0.65, 0.70, 0.75	0.423	0.701
	50	350		0.286	0.473
	100	700		0.198	0.327
	200	1400		0.139	0.229
	400	2800		0.098	0.161
8	25	200	0.25, 0.30, 0.35, 0.40, 0.60, 0.65, 0.70, 0.75	0.392	0.647
	50	400		0.266	0.439
	100	800		0.185	0.304
	200	1600		0.129	0.213
	400	3200		0.091	0.150
9	25	225	0.25, 0.30, 0.35, 0.40, 0.45, 0.60, 0.65, 0.70, 0.75	0.368	0.606
	50	450		0.250	0.410
	100	900		0.173	0.284
	200	1800		0.121	0.199
	400	3600		0.085	0.140
10	25	250	0.25, 0.30, 0.35, 0.40, 0.45, 0.55, 0.60, 0.65, 0.70, 0.75	0.347	0.570
	50	5000		0.236	0.386
	100	1000		0.164	0.268
	200	2000		0.115	0.188
	400	4000		0.081	0.132

(Continued)

Table 5.1 (Continued) **Comparison of Designs for Estimation of an LD_{50} and LD_{90}**

Doses	Cell	Sample	P_0[a]	L^{*}[b] Probit	L^{*}[b] Logit
			LD_{90}		
5	25	125	0.10, 0.80, 0.85, 0.90, 0.95	0.729	1.426
	50	250		0.500	0.966
	100	500		0.348	0.670
	200	1000		0.244	0.469
	400	2000		0.172	0.330
6	25	150	0.10, 0.75, 0.80, 0.85, 0.90, 0.95	0.668	1.298
	50	300		0.457	0.877
	100	600		0.318	0.607
	200	1200		0.223	0.425
	400	2400		0.157	0.299
7	25	175	0.05, 0.70, 0.75, 0.80, 0.85, 0.90, 0.95	0.633	1.259
	50	350		0.432	0.842
	100	700		0.300	0.580
	200	1400		0.210	0.405
	400	2800		0.148	0.284
8	25	200	0.05, 0.10, 0.70, 0.75, 0.80, 0.85, 0.90, 0.95	0.602	1.148
	50	400		0.417	0.788
	100	800		0.292	0.550
	200	1600		0.205	0.386
	400	3200		0.145	0.272
9	25	225	0.05, 0.10, 0.15, 0.70, 0.75, 0.80, 0.85, 0.90, 0.95	0.595	1.121
	50	450		0.414	0.776
	100	900		0.290	0.543
	200	1800		0.205	0.382
	400	3600		0.144	0.269
10	25	250	0.05, 0.10, 0.15, 0.65, 0.70, 0.80, 0.85, 0.90, 0.95	0.569	1.070
	50	500		0.396	0.740
	100	1000		0.278	0.518
	200	2000		0.196	0.364
	400	4000		0.138	0.257

[a] P_0 is the probability of response.
[b] L^{*} is the measurement of precision.

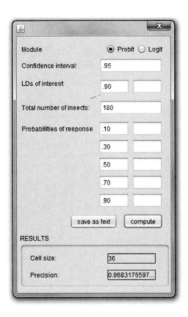

Figure 5.1   Start screen for OptiDose program.

To answer the first question, Dr. Maven assumes a basic design with five doses, a total sample of $n = 180$ and equal numbers of caterpillars at each dose. She uses three different spacings for probabilities of response. The first is an even spacing between 0.10 and 0.90 (i.e., 10% and 90% response). The second is placement of all responses in the upper half of the probability scale. The third is an intermediate design in which one probability is low and the other probabilities are in the upper portion of the scale. OptiDose gives the following estimate of L^*, the measurement of precision based on the width of the 95% confidence limits used to determine the precision of the LD estimate:

Probabilities of Response					L^* for LD_{90}
0.10	0.30	0.50	0.70	0.90	0.87
0.50	0.60	0.70	0.80	0.90	1.09
0.10	0.60	0.70	0.80	0.90	0.71

Based on this table, Dr. Maven selects a design with most of the doses causing around 90% response and one or two causing <50% response, giving a precise estimate of the LD_{90}.

Next, Dr. Maven explores the effect of varying the number of doses, again with $n = 180$ and equal numbers treated with each dose:

Probability of Response							L^* for LD_{90}
0.05	0.90	0.95					0.66
0.05	0.80	0.85	0.90	0.95			0.59
0.05	0.10	0.15	0.70	0.75	0.80	0.90	0.95	0.67

Her best choice seems to be to use five application rates as it has the most precise estimate (i.e., 0.59). There are numerous ways to distribute the 180 caterpillars among five application rates. Dr. Maven compares an equal allocation ($n = 36$), assignment of more insects to the upper and lower extremes of response, and assignment of more insects to the intermediate probabilities of response. For the best design according to the first two questions, OptiDose shows these results for sample allocation:

Probability of Response					L^* for LD_{90}
0.05	0.80	0.85	0.90	0.95	
36	36	36	36	36	0.59
54	24	24	24	54	0.63
18	48	48	48	18	0.59

The first and third allocation schemes offer equal precision. Dr. Maven also is interested in LD_{50} and LD_{95}, so she examines the relative precision of the two designs for estimation of these response levels:

Probability of Response					L^* for LD_{95}	L^* for LD_{50}
0.05	0.80	0.85	0.90	0.95		
36	36	36	36		0.71	0.66
18	48	48	48	18	0.78	0.88

Equal allocation clearly results in greater precision for both LD_{95} and LD_{50}, so Dr. Maven now knows how to design her bioassay. She will place her application rates so that one is at a low probability of response and the others are located so that they produce mortalities between 80% and 95%. She will test an equal number of caterpillars at each application rate.

The final step in the bioassay involves use of the preliminary results to estimate the application rates that should produce the responses for the experimental design. Use of another computer program such as PoloJR will provide these starting levels for the first replication of the bioassay.

5.1.2 Number of Doses

Suppose we want to examine a greater number of probabilities and dose placement, but we do not want to spend days on the best dose estimates. OptiDose
provides general guidance concerning the number of doses to use in a bioassay,
but it was not designed specifically to study the number that would be theoretically
optimal. Müller and Schmitt[19] addressed the problem of how to select the number
of doses to provide the best estimate of LD_{50} based on the criterion of optimizing
the asymptotic variance of LC_{50}. They concluded that use of as many doses as
possible should be tested in a bioassay and that use of only three doses cannot be
recommended.

These results may indicate minimum requirements for a basic bioassay, but further
investigations are clearly necessary to explore other possible combinations of sample
size and dose placement applicable to other LDs in either basic or specialized binary
bioassays.

5.2 OPTIDOSE STATISTICS

After repeated requests for a new Polo program that will work on Macintosh
OS and every biyearly update of Windows software, a new program was designed
specifically for the study of dose placement and sample size for any dose that would
cause >0% and <100% mortality. OptiDose[20] calculates the length of the confidence
interval (L) for μ_0, the dose level that produces a certain probability of response
P_0, where $0 < P_0 < 1$. In this program, P_0 is estimated to 35 digits, thus permitting
examination of doses necessary to estimate the probit and logit 9. The equations
used in OptiDose are based on those given by Smith and Robertson,[18] but with the
following substitutions: L^* was replaced with L, v^* with v, g^* with g, and w_t^* with
w_t. There are T dose levels, and at the tth dose level, x_t is the dose and n_t is the
number of subjects. The probability of response is $P_t = F(\beta_1 + \beta_2 x_t)$, where β_1 and
β_2 are unknown parameters, and P_0 is the probability of response associated with
the LD of interest. $F(\bullet)$ is a cumulative distribution function (CDF) with density
$f(\bullet)$ and inverse $F^{-1}(\bullet)$.

The standard normal (Gaussian) CDF is

$$F(x) = \frac{1}{\sqrt{2\pi}} \int_{-\infty}^{x} e^{-\frac{1}{2}z^2} \, dz.$$

The logistic CDF is

$$F(x) = 1/(1 + e^{-x})$$

$$w_t = f\left(F^{-1}(P_t)\right)^2 / P_t(1 - P_t).$$

Covariances of β_1 and β_2 are

$$\begin{bmatrix} v_{11} & v_{12} \\ v_{12} & v_{22} \end{bmatrix} = \begin{bmatrix} \sum n_t w_t & \sum n_t w_t F^{-1}(P_t) \\ \sum n_t w_t F^{-1}(P_t) & \sum n_t w_t \left[F^{-1}(P_t)\right]^2 \end{bmatrix}^{-1}.$$

z is the abscissa of the normal probability distribution corresponding to the significance level, which is 5% in OptiDose.

$$g = z^2 v_{22}$$

The measure of precision is

$$L = \frac{2z}{1-g} \sqrt{v_{11} + 2F^{-1}(P_0)\, v_{12} + \left[F^{-1}(P_0)\right]^2 v_{22} - g\left(v_{11} - v_{12}^2 / v_{22}\right)}.$$

5.3 BASIC BINARY BIOASSAYS

Dr. Maven uses values of the probit L^* to assess the quality of each bioassay design. The estimated 95% confidence intervals for both the probit and logit models with cell sizes of 25, 50, 100, 200, and 400 with 5–10 doses clearly show that estimates become more precise as total sample size increases (see Table 5.1). Likewise, precision increases as the numbers of doses increases. Smith and Robertson[18] previously concluded that minimally acceptable precision is achieved with 240 test subjects, which had a logit L^* (Robertson and Preisler[21]) of 0.58. This value corresponds with a probit L^* of ca. 0.34. Suspecting that minimally acceptable might be a euphemism for marginal precision, Dr. Maven decides that a probit L^* value ≤ 0.30 should indicate fully acceptable precision. She does not want to base future decisions about either research or management of the Godfather populations on results that could have been more precise if more insects or doses were tested.

In an ideal basic bioassay for comparisons at LD_{50}, Paula concludes that Jessica should test five doses with at least 100 insects per dose; with 6–10 doses, she should test at least 50 insects per dose. Bioassays to estimate the LD_{90} require twice as many test subjects. For fully acceptable precision estimation of LD_{50} and LD_{90}s, minimum sample sizes of 300–500 and 600–1000 are required, respectively.

For simultaneous estimation of both LD_{50} and LD_{90}, neither estimate can ever be based on extrapolation. One solution is to use at least one dose that causes approximately 10% mortality, and the remainder of the doses between 20% and 95% mortality. For example, results in Table 5.2 show probit $L*$ values for various designs derived from the results from designs shown in Table 5.1. With five doses, the design with 500 test subjects (100/cell) provides marginal precision, and ≥750 test subjects ensures acceptable precision. For the six doses shown, 600 test subjects (100/cell) appear to provide fully acceptable precision. When 7–12 doses are tested, most estimates of LD_{50} are precise regardless of dose placement or sample size, but LD_{90} is less precise unless large sample sizes are used. The practical problems associated with simultaneous precise estimation will be discussed in Section 5.5.

5.4 SPECIALIZED BINARY BIOASSAYS

Estimation of extreme levels such as LD_{95} (Table 5.3) and LD_{99} (Table 5.4) requires more doses and larger sample sizes than a basic bioassay does. For estimation of LD_{95}, sample sizes ≥1000 will provide fully acceptable precision with 5–10 doses. For these optimal designs, one to two doses cause ≤80% mortality, and the remainder (~six or greater) cause mortality from 90% to 99% mortality. A specialized bioassay to estimate an LD_{95} might be useful in studies of resistance such as genetic patterns, inheritance, and surveys of occurrence (see Chapter 9).

For precise estimation of the LD_{99}, none of the designs with five to six doses is acceptable. Use of at least seven doses is necessary, but total sample sizes are 3000–3600 based on doses expressed as base 10 percentages. One to three doses must cause ≤0.80% mortality. The rest (~6 or greater) caused between 97% and 99.99995% mortality. At this extreme dose, finding the exact dose that causes the upper level of mortality in the series will require that 99,995 of 100,000 test subjects die. Thus, a sample size of 100,000 will be necessary just to identify the upper dose in the series of dilutions tested. The total sample size required for the replicated experiment is completely impractical under most circumstances for most species.

5.5 PRACTICAL CONSIDERATIONS

5.5.1 Basic Bioassays

Despite all of her reading, Paula Maven has never found a detailed discussion of the practical aspects of doing a typical bioassay such as one that is done when a new insect species is first received in a laboratory, or when a new chemical or biological or physical treatment is to be tested on a species for which bioassays have been done before. Often, the number of test subjects available for the bioassay is the primary limiting factor for how many doses can be tested. As with Dr. Maven's initial shipments of Godfather larvae from field populations, fewer than 500 insects may constitute the entire sample. Table 5.5 shows the precision of designs for total sample sizes of 250, 300, 400, and 500 test subjects. These results indicate that, if possible, a total sample size of 500 test subjects with eight doses (63 insects/dose) should be tested. Jessica finally has her guidelines!

Besides limits on available test subjects, a standardized test such as a population assessment or a screening experiment is usually done under severe time constraints (Around this time Jessica has a deer-in-headlights look to her). To minimize the time involved, a standard procedure based on preliminary data can be used to select the doses for the first replication of the bioassay. This part of the experiment, the *dose-fixing* phase, consists of using a broad series of logarithmic dilutions (e.g., 0.001, 0.003, 0.01, 0.03, 0.1, 0.3, 1.0 ppm, etc.) to test about 10 insects per dose. Results of dose-fixing preliminary research can identify the narrower range of effective doses based on mortality that occurs a short time after treatment. A series of dilutions between the doses that very crudely estimate 1%–99% mortality can also be identified from these preliminary results by analyzing that data with the PoloJR program using the "Graphics" option. Once the effective dose ranges are identified for each population, five to seven doses can be selected and tested in the first actual replication. After the first replication is completed, the doses can be adjusted further in the second and third replications. These doses will then be tested in the basic bioassay. To evenly test each dose as these adjustments are made, it is important to continue testing doses, even with the smallest number of insects (~five), while new doses are being developed and tested. No more than one dose should cause 100% mortality, or a plateau will form rather than a line. As always, the experiment should be replicated a minimum of three times, with a minimum of five doses in each replication.

5.5.2 Specialized Bioassays

At least two dose-fixing experiments are usually required to determine doses for a specialized bioassay. The first is the same type done for the general bioassay: a logarithmic series of dilutions must be done to tentatively identify the effective range for 10%–95% mortality. In this series, a tentative LD_{5-80} and doses in the upper range (90%–99% mortality) should be identified. When the specialized bioassay is done to estimate LD_{95}, 5–10 doses should then be tested with about 50 test subjects per dose. In the next replication, doses can be adjusted further to achieve 90%–99% mortality. When the specialized bioassay is done to estimate LD_{99} (i.e., diagnostic dose, discriminating dose), two dose-fixing experiments should be done, followed by selection of 7–10 doses that cause 97%–99.99995% mortality. Each of the final doses should then be tested as many times as necessary to achieve the total required sample of >700,000 test subjects. In reality, if a population is collected from a location that has tolerant individuals (e.g., spray failures have been present), necessary sampling number may be much smaller. In addition, if tolerance in a population is not present in a high percentage, the diagnostic dose may be lowered (i.e., 95%) to reduce the sample number of necessary test subjects.

Table 5.2 Designs for Simultaneous Estimation of the LD_{50} and LD_{90} Assuming the Probit Model

ND	P_L (Range)	P_0	NC	NS	L*
5	0.01–0.80	0.90, 0.95, 0.975, 0.99	100	500	0.456–0.480
			150	750	0.370–0.377
			175	875	0.342–0.346
			200	1000	0.319–0.321
			300	1500	0.257–0.260
			400	2000	0.221–0.225
			500	2500	0.197–0.201
			600	3000	0.179–0.183
			700	3500	0.165–0.170
			800	4000	0.154–0.159
			900	4500	0.145–0.149
			1000	5000	0.138–0.142
6		0.90, 0.925, 0.95, 0.975, 0.99	100	600	0.395–0.436
			175	1050	0.295–0.314
			185	1110	0.288–0.304
			200	1200	0.276–0.292
			300	1800	0.225–0.234
			400	2400	0.195–0.201
			500	3000	0.174–0.179
			600	3600	0.159–0.163
			700	4200	0.147–0.150
			800	4800	0.137–0.140
			900	5400	0.129–0.132
			1000	6000	0.123–0.125
7		0.90, 0.925, 0.95, 0.975, 0.98, 0.99	100	700	0.368–0.380
			150	1050	0.299–0.302
			175	1225	0.276–0.277
			200	1440	0.257–0.258
			300	2100	0.208–0.210
			400	2800	0.179–0.182
			500	3500	0.159–0.162
			600	4200	0.145–0.148
			700	4900	0.134–0.137
			800	5600	0.125–0.128
			900	6300	0.118–0.121
			1000	7000	0.112–0.115

(*Continued*)

Table 5.2 (Continued)   **Designs for Simultaneous Estimation of the LD$_{50}$ and LD$_{90}$ Assuming the Probit Model**

ND	P$_L$ (Range)	P$_0$	NC	NS	L*
8	0.90, 0.925, 0.95, 0.96, 0.975, 0.98, 0.99		100	800	0.337–0.346
			125	1000	0.301–0.305
			150	1200	0.274–0.276
			175	1400	0.253–0.254
			200	1600	0.236–0.237
			300	2400	0.190–0.194
			400	3200	0.164–0.167
			500	4000	0.146–0.150
			600	4800	0.133–0.137
			700	5600	0.123–0.127
			800	6400	0.115–0.118
			900	7200	0.108–0.112
			1000	8000	0.103–0.106
9	0.90, 0.925, 0.93, 0.95, 0.96, 0.975, 0.98, 0.99		85	765	0.338–0.358
			100	900	0.310–0.325
			200	1800	0.218–0.222
			300	2700	0.178–0.179
			400	3600	0.154–0.155
			500	4500	0.137–0.138
			600	5400	0.125–0.126
			700	6300	0.116–0.117
			800	7200	0.108–0.109
			900	8100	0.102–0.103
			1000	9000	0.096–0.097
10	0.90, 0.915, 0.925, 0.93, 0.95, 0.96, 0.975, 0.98, 0.99		85	850	0.316–0.343
			100	1000	0.289–0.312
			200	2000	0.203–0.212
			300	3000	0.165–0.171
			400	4000	0.143–0.148
			500	5000	0.128–0.131
			600	6000	0.117–0.120
			700	7000	0.108–0.111
			800	8000	0.101–0.103
			900	9000	0.095–0.099
			1000	10000	0.090–0.092

Note: Column headings are number of doses (*ND*), probabilities in the lower range of response (*P*$_L$), probability of response (*P*$_0$), number of cells (*NC*), total number in the sample (*NS*), and measurement of precision (*L**).

Table 5.3 Optimal Designs for Precise Estimation of the LD_{95}

ND	P_L (Range)	P_0	NC	NS	L*
5	0.01–0.80	0.90, 0.95, 0.975, 0.99	100	500	0.456–0.480
			150	750	0.370–0.377
			175	875	0.342–0.346
			200	1000	0.319–0.321
			300	1500	0.257–0.260
			400	2000	0.221–0.225
			500	2500	0.197–0.201
			600	3000	0.179–0.183
			700	3500	0.165–0.170
			800	4000	0.154–0.159
			900	4500	0.145–0.149
			1000	5000	0.138–0.142
6		0.90, 0.925, 0.95, 0.975, 0.99	100	600	0.395–0.436
			175	1050	0.295–0.314
			185	1110	0.288–0.304
			200	1200	0.276–0.292
			300	1800	0.225–0.234
			400	2400	0.195–0.201
			500	3000	0.174–0.179
			600	3600	0.159–0.163
			700	4200	0.147–0.150
			800	4800	0.137–0.140
			900	5400	0.129–0.132
			1000	6000	0.123–0.125
7		0.90, 0.925, 0.95, 0.975, 0.98, 0.99	100	700	0.368–0.380
			150	1050	0.299–0.302
			175	1225	0.276–0.277
			200	1440	0.257–0.258
			300	2100	0.208–0.210
			400	2800	0.179–0.182
			500	3500	0.159–0.162
			600	4200	0.145–0.148
			700	4900	0.134–0.137
			800	5600	0.125–0.128
			900	6300	0.118–0.121
			1000	7000	0.112–0.115

(*Continued*)

Table 5.3 (Continued)   Optimal Designs for Precise Estimation of the LD$_{95}$

ND	P_L (Range)	P_0	NC	NS	L*
8	0.90, 0.925, 0.95, 0.96, 0.975, 0.98, 0.99		100	800	0.337–0.346
			125	1000	0.301–0.305
			150	1200	0.274–0.276
			175	1400	0.253–0.254
			200	1600	0.236–0.237
			300	2400	0.190–0.194
			400	3200	0.164–0.167
			500	4000	0.146–0.150
			600	4800	0.133–0.137
			700	5600	0.123–0.127
			800	6400	0.115–0.118
			900	7200	0.108–0.112
			1000	8000	0.103–0.106
9	0.90, 0.925, 0.93, 0.95, 0.96, 0.975, 0.98, 0.99		85	765	0.338–0.358
			100	900	0.310–0.325
			200	1800	0.218–0.222
			300	2700	0.178–0.179
			400	3600	0.154–0.155
			500	4500	0.137–0.138
			600	5400	0.125–0.126
			700	6300	0.116–0.117
			800	7200	0.108–0.109
			900	8100	0.102–0.103
			1000	9000	0.096–0.097
10	0.90, 0.915, 0.925, 0.93, 0.95, 0.96, 0.975, 0.98, 0.99		85	850	0.316–0.343
			100	1000	0.289–0.312
			200	2000	0.203–0.212
			300	3000	0.165–0.171
			400	4000	0.143–0.148
			500	5000	0.128–0.131
			600	6000	0.117–0.120
			700	7000	0.108–0.111
			800	8000	0.101–0.103
			900	9000	0.095–0.099
			1000	10000	0.090–0.092

Note: Column headings are number of doses (*ND*), probabilities in the lower range of response (P_L), probability of response (P_0), number of cells (*NC*), total number in the sample (*NS*), and measurement of precision (*L**).

Table 5.4   Optimal Designs for Precise Estimation of the LD_{99}

ND	P_L	P_0	NC	NS	L^*
7	0.01–0.80	0.97, 0.98, 0.99, 0.995, 0.9995, 0.99995	300	2100	0.372–0.421
			400	2800	0.327–0.362
			500	3500	0.293–0.323
			600	4200	0.267–0.294
			700	4900	0.247–0.271
			800	5600	0.231–0.253
			900	6300	0.218–0.239
			1000	7000	0.206–0.226
8		0.97, 0.98, 0.99, 0.9925, 0.995, 0.9995, 0.99995	200	1600	0.430–0.471
			300	2400	0.350–0.380
			400	3200	0.303–0.327
			500	4000	0.271–0.292
			600	4800	0.247–0.265
			700	5600	0.224–0.245
			800	6400	0.214–0.229
			900	7200	0.201–0.216
			1000	8000	0.191–0.205
9		0.97, 0.975, 0.98, 0.99, 0.9925, 0.995, 0.9995, 0.99995	100	900	0.545–0.627
			200	1800	0.382–0.428
			300	2700	0.311–0.346
			400	3600	0.269–0.298
			500	4500	0.240–0.266
			600	5400	0.219–0.242
			700	6300	0.203–0.224
			800	7200	0.190–0.209
			900	8100	0.179–0.197
			1000	9000	0.170–0.187
10		0.97, 0.975, 0.98, 0.985, 0.99, 0.9925, 0.995, 0.9995, 0.99995	100	1000	0.505–0.577
			200	2000	0.354–0.395
			300	3000	0.288–0.319
			400	4000	0.249–0.275
			500	5000	0.223–0.245
			600	6000	0.203–0.224
			700	7000	0.188–0.207
			800	8000	0.176–0.193
			900	9000	0.166–0.193
			1000	10000	0.157–0.172
			1500	15000	0.128–0.141
			2000	20000	0.111–0.122

Note: Column headings are number of doses (*ND*), probabilities in the lower range of response (*P_L*), probability of response (*P_0*), number of cells (*NC*), total number in the sample (*NS*), and measurement of precision (*L^**).

Table 5.5 Precision of Bioassays with Sample Sizes of 250–500

N	NC	D	P_0	L_{50}^*	L_{90}^*
250	50	5	0.05, 0.80, 0.85, 0.90, 0.95	0.558	0.497
	42	6	0.05, 0.35, 0.80, 0.85, 0.90, 0.95	0.452	0.538
	31	8	0.10, 0.25, 0.35, 0.75, 0.80, 0.85, 0.90, 0.95,	0.404	0.582
	25	10	0.05, 0.10, 0.25, 0.30, 0.70, 0.80, 0.85, 0.90, 0.95	0.396	0.597
	21	12	0.05, 0.25, 0.30, 0.35, 0.65, 0.70, 0.75, 0.80, 0.85, 0.90, 0.95, 0.99	0.407	0.615
300	60	5	0.05, 0.80, 0.85, 0.90, 0.95	0.506	0.452
	50	6	0.05, 0.35, 0.80, 0.85, 0.90, 0.95	0.412	0.490
	38	8	0.10, 0.25, 0.30, 0.75, 0.80, 0.85, 0.90, 0.95	0.362	0.522
	30	10	0.05, 0.10, 0.25, 0.30, 0.70, 0.75, 0.80, 0.85, 0.90, 0.95	0.360	0.542
	25	12	0.05, 0.25, 0.30, 0.35, 0.65, 0.70, 0.75, 0.80, 0.85, 0.90, 0.95, 0.99	0.372	0.560
400	80	5	0.05, 0.80, 0.85, 0.90, 0.95	0.435	0.389
	67	6	0.05, 0.35, 0.80, 0.85, 0.90, 0.95	0.354	0.421
	50	8	0.10, 0.25, 0.35, 0.75, 0.80, 0.85, 0.90, 0.95	0.315	0.452
	40	10	0.05, 0.10, 0.25, 0.35, 0.70, 0.80, 0.85, 0.90, 0.95	0.311	0.466
	33	12	0.05, 0.25, 0.30, 0.35, 0.65, 0.70, 0.75, 0.80, 0.85, 0.90, 0.95, 0.99	0.322	0.483
500	100	5	0.05, 0.80, 0.85, 0.90, 0.95	0.388	0.346
	83	6	0.05, 0.35, 0.80, 0.85, 0.90, 0.95	0.317	0.377
	63	8	0.10, 0.25, 0.35, 0.75, 0.80, 0.85, 0.90, 0.95	0.280	0.352
	50	10	0.05, 0.25, 0.30, 0.65, 0.70, 0.75, 0.80, 0.85, 0.90, 0.95	0.286	0.409
	42	12	0.05, 0.25, 0.30, 0.35, 0.65, 0.70, 0.75, 0.80, 0.85, 0.90, 0.95, 0.99	0.284	0.426

Note: Column headings are number of test subjects (*N*), number in each cell of cells (*NC*), and number of doses (*D*); P_0 is the ideal probability of response at each dose, and L^* is the measurement of precision at the 50% or 90% response levels.

5.6 REALITY CHECKLIST FOR BIOASSAYS

The various sample sizes and the way doses must be placed for precise estimation of LD_{50} and LD_{90} are fairly flexible compared with the stringent requirements for estimation of an LD_{99}. The requirements for estimation of LD_{99} are complicated and the numbers of test subjects are huge. Historically, entomologists have tended to make carefree recommendations about LDs that should be estimated for specific purposes such as resistance testing to separate different genotypes, without realizing the numbers of test subjects that would be required to estimate that LD or how the doses would need to be placed for purposes of precise estimation. These types of recommendations have been intended to make bioassays definitive but have instead made them impossible to do with any degree of precision. It is simply impossible to have a precise LD_{99} developed with only 1000 tested or better yet to confirm resistance in field populations without testing tens of thousands of individuals. More recently, the idea of testing $LD_{99} \times 2$ dose on a population to screen for resistance has been suggested. The idea of using a discriminating dose is to screen for individuals that are tolerant to a pesticide. Using a dose twice as potent as the original will leave no individuals remaining in the population and will defeat the purpose of using the discriminating dose in the first place. In fact, using such a high dose will most likely kill all individuals tested, making it appear that no tolerance exists for a chemical. Hogwash.

Sample sizes are limited by the realities of insect production or collection even on a mass scale. If arthropods were like bacteria and each species could be easily reared in huge numbers in a small area, numbers of test subjects would not be a concern. However, rearing of arthropods is an activity that usually requires large areas of laboratory space *if* the species can be reared at all. Of more than one million species that have been identified, fewer than 1500 arthropod species have been reared continuously and on a large scale.[22] With Lepidoptera such as Dr. Maven's Godfather larvae, collections of population samples and subsamples of laboratory populations are extremely time and labor intensive. Theoretical requirements cannot be changed, but the type of bioassays used with general groups of arthropods can certainly be adjusted to achieve the best possible precision.

So, what types of bioassays should Dr. Paula Maven consider doing on her Godfather larvae considering her small lab space? Like most Lepidoptera, vast numbers of insects are not available even with a well-established and highly efficient rearing operation. For most experiments with Lepidoptera, 300–500 insects are available for testing at one time. For experiments in which comparisons to field populations are involved—screening, population responses, natural variation in response—a basic design for estimation of the LD_{50} will suffice and comparisons at the LD_{90} should be avoided unless >500 test subjects of each population are available. Both LD_{50} and LD_{90} should be estimated for comparisons of relative effectiveness only when >500–750 test subjects of each population are available. These high numbers usually occur with organisms such as ticks and many species of Diptera; these organisms are generally available in large numbers even in laboratory situations.

5.7 CONCLUSIONS

The key to a reliable binary quantal response bioassay is good experimental design. A well-designed experiment includes adequate sample sizes, complete replication, and doses carefully placed to estimate the LD or concentration of interest. Use of OptiDose provides guidance both to adequate sample size and to good placement of doses.

REFERENCES

1. Robertson, J. L., Smith, K. C., Savin, N. E., and Lavigne, R. J., Effects of dose selection and sample size on precision of lethal dose estimates in dose-mortality regression, *J. Econ. Entomol.* 77, 833, 1984.

2. Finney, D. J., Probit Analysis, Cambridge University Press, Cambridge, England, 1971.

3. Robertson, J. L., Toxicity of Zectran aerosol to the California oakworm, a primary parasite, and a hyperparasite, *Environ. Entomol.* 1, 1, 115–117, 1972.

4. Robertson, J. L., Preisler, H. K., Ng, S. S., Hickle, L. A., and Gelernter, W. D., Natural variation: A complicating factor in bioassays with chemical and microbial insecticides, *J. Econ. Entomol.* 88, 1, 1995.

5. Robertson, J. L., Boelter, L. M., Russell, R. M., and Savin, N. E., Variation in response to insecticides by Douglas-fir tussock moth, Orgyia pseudotsugata (Lepidoptera: Lymantriidae), populations, *Can. Entomol.* 110, 325, 1978.

6. Robertson, J. L. and Rappaport, N. G., Direct, indirect, and residual toxicities of insecticide sprays to western spruce budworm, Choristoneura occidentalis (Lepidoptera: Tortricidae), *Can. Entomol.* 111, 1219, 1979.

7. Robertson, J. L., Gillette, N. L., Lucas, B. A., Savin, N. E., and Russell, R. M., Comparative toxicity of insecticides to *Choristoneura* species (Lepidoptera: Tortricidae), *Can. Entomol.* 110, 399, 1978.

8. Bergh, J. C., Rugg, D., Jansson, R. K., McCoy, C. W., and Robertson, J. L., Monitoring the susceptibility of citrus rust mite (Acari: Eriophyidae) populations to abamection, *J. Econ. Entomol.* 92, 781, 1999.

9. Robertson, J. L., Gillette, N. L., Look, M., Lucas, B. A., and Lyon, R. L., Toxicity of selected insecticides to western spruce budworm, *J. Econ. Entomol.* 69, 99, 1976.

10. Dowd, P. F. and Sparks, T. C., Relative toxicity and ester hydrolysis of pyrethroids in the soybean looper and tobacco budworm, *J. Econ. Entomol.* 81, 1014, 1988.

11. Prabhaker, N., Coudriet, D. L., and Toscano, N. C., Effect of synergisis on organophosphate and permethnn resistance in sweetpotato whitefly (Homoptera: Aleyrodidae), *J. Econ. Entomol.* 81, 34, 1988.

12. Payne, G. T, Blenk, R. C., and Brown, T. M., Inheritance of permethrin resistance in the tobacco budworm (Lepidoptera: Noctuidae), *J. Econ. Entomol.* 81, 65, 1988.

13. Pree, D. J., Inheritance and management of cyhexatin and dicofol resistance in the European red mite (Acari: Tetranychidae), *J. Econ. Entomol.* 80, 1106, 1987.

14. Brown, B. W. Jr., Planning a quantal assay of potency, *Biometrics* 22, 322, 1966.

15. Freeman, P. R., Optimum Bayesian sequential estimation of the median effective dose, *Biometrika* 57, 79, 1970.

16. Tsutakawa, R. K., Selection of dose levels for estimating a percentage point of a logistic quantal response curve, *Appl. Stat.* 29, 25, 1980.

17. Haverty, J. I. and Robertson, J. L. Laboratory bioassays for selecting candidate insecticides and application rates for field tests on the western spruce budworm, *J. Econ. Entomol.* 75, 179, 1982.

18. Smith, K. C. and Robertson, J. L., DOSESCREEN: A computer program to aid dose placement, *USDA Forest Service Gen. Tech. Rep.,* PSW-78, 1984.

19. Müller, H.-G. and Schmitt, T., Choice of number of doses for maximum likelihood estimation of the ED_{50} for quantal dose–response data, *Biometrics* 6, 117, 1990.

20. LeOra Software. *OptiDose*. PO Box 562, Parma, MO 63870. See http://www.LeOra Software.com.
21. Robertson, J. L. and Preisler, H. K., *Pesticide Bioassays with Arthropods*, CRC Press, Boca Raton, FL, 1992.
22. Singh, P. and Moore, R. F., *Handbook of Insect Rearing, Vols. I and II*, Elsevier, Amsterdam, 1985.

CHAPTER 6

Natural Variation in Response

As university research goes, Paula spends more and more time writing research proposals and quarterly reports, giving presentations, writing peer-reviewed publications, and schmoozing with local growers and commodity groups. In order to balance all these tasks, Paula and Jessica develop a new system. Paula writes the protocols, does the data analyses, and writes, while Jessica completes the bioassays, manages the field crew, does the field assessments, orders supplies, and keeps everything organized. Or, as Jessica puts it, she does all the work and Paula thinks a lot. Once the laboratory colony has been well established (after two rough starts, once with no eggs hatching and then with no pupae emerging) Jessica does repetitive bioassays with three chemicals representative of different chemical classes and a standard preparation of the microbial toxicant *Bacillus thuringiensis* subsp. *kurstaki*. Paula wants to monitor the quality of the insect colony with these tests.

After six bioassays with each chemical, Paula becomes concerned and calls Jessica into her office to talk about the results thus far. Something is causing the results to vary from one bioassay to another, and Paula is freaking out about the quality of the insect colony (is it crashing again?), the purity of formulated samples (where did she order them?), the cleanliness of the lab (is there mold contamination in the lab?), and everything else including the number of sunspots that occurred during the bioassays. In other words, she has become frantic, irrational, and a raving lunatic.

"I don't understand what's going on here, Jessica. Are your pipettes calibrated? Perhaps, you are not measuring the amounts of the chemicals the same when you make up new solutions? Are you making a fresh batch of formulations each time you do a bioassay? Is it possible..." Jessica stops Paula midsentence to reassure her that, yes, everything is done exactly the same way every time: "Look, we even store the samples at −40°. The volumes are always the same—that's why you bought all new expensive pipetters remember? Now what else do you want me to do?" There's a pause. "We will fix this. It is just another weekly problem to solve." So, Paula finds her old notebook from her bioassay course plus her copy of Finney's *Statistical Methods in Biological Assay*.[1] The very sight of the equations is like electroshock therapy, and she begins reading about "Standard Curve Estimation." Here she finds the answer to the problem: "In general, the assumption that a response curve once determined can be used in future assays is inadmissible".[1] Now she also remembers the corollary of Robertson et al.[2]: "Because of natural variation, responses of groups of insects tested at any one time will therefore never be exactly the same as responses of another group tested either at the same time or a different time, regardless of the extent to which bioassay techniques are standardized."

6.1 DEFINITION

Natural variation is best defined as a numerical difference in response that is detected each time a bioassay is repeated with one genetic strain, either within a single generation or more than one generation (Robertson et al.[2]). Such variation is akin to background noise and can never be eliminated: it is another reminder that each bioassay provides an estimate of response, not a measurement.

6.2 STATISTICAL BOUNDARIES OF NATURAL VARIATION

For individual data sets, probit or other regression techniques capture the variation among data points. Standard errors and 95% confidence limits (CL) for individual lines thus do not define natural variation among data sets. Instead, repeated bioassays are essential. As more and more bioassays are done, estimates of natural variation become more and more reliable. Either lethal doses (LDs) or LD ratios can be used to determine the boundaries.

Based on large sample theory and assuming that the samples in all bioassays provide a random sample of responses of n present and future bioassays, the approximate 95% limits of natural variation are mean ± 1.96 (SD) where mean and SD are the mean and standard deviation of the LDs from n samples. When LD ratios are used, as in bioassays with microbial preparations, the equation is the same, with mean and SD calculated from the lethal times from the n samples. Dr. Maven next considers some possible sources of natural variation.

6.3 LEVELS OF VARIATION

6.3.1 Sibling Groups

The smallest unit in population formation is the egg mass, which Dr. Maven calls the *sibling group*. Information about the quantal responses of sibling groups can be difficult to obtain because of limitations in sample sizes (i.e., total number of eggs laid) and demographic factors, such as percentage hatch and subsequent survival. For example, Robertson (unpublished data) tested western spruce budworm (*Choristoneura occidentalis* Freeman) and green budworm (*C. viridis* Freeman) that were reared from single egg masses and isolated throughout their development. As sixth instars, the insects were tested with either acephate or carbaryl by topical

Table 6.1   Results of Hypotheses Tests for Parallelism or Equality for *C. occidentalis* Freeman and *C. viridis* Freeman Sibling Groups

Chemical	Species	Groups Tested	Equality	Parallelism
Acephate	C.o	34	R	R
	C.v	12	NR	NR
Carbaryl	C.o	15	R	NR
	C.v	11	R	NR

application. Despite the small total sample sizes available for each regression, results of hypotheses tests indicated that the hypothesis of equality was rejected except when *C. viridis* sibling groups were exposed to acephate (see Table 6.1). In contrast, the hypothesis of parallelism was rejected only when *C. occidentalis* was treated with carbaryl. These results may indicate the incremental contribution of each sibling group to the natural variation among cohorts within a generation.

6.3.2 Cohorts within a Single Generation

Egg masses hatch asynchronously so that larvae from different sibling groups are present within a given developmental stage and different developmental stages may be present at a given time during population development (e.g., Price[3] and Morris and Miller[4]). One of the reasons that Paula and Jessica established a laboratory colony of Godfather larvae was to ensure a ready supply of insects of uniform age or size for testing on a year-round basis. They used mass-production methods common to many rearing facilities; these methods include provision of the insects with an overabundance of food, exclusion of predators and parasites, and manipulation of development by storage of egg masses for weekly infusion into the colony. The end result is a population with cohorts defined by time and with generations that can be precisely delimited. These factors permit definition of natural variation without interference from outside factors that would also cause responses to vary.

Results of repeated bioassays suggest the extent to which natural variation within strains of arthropod species reared even under highly controlled conditions can complicate detection of other differences related to pesticide resistance, varietal differences of host plants, or product quality. However, Paula remembers that 5% of the values or ratios will always fall outside the limits estimated by the equation. If she observes a trend in decreasing response in repeated bioassays, then tentative conclusions about resistance and other biochemical or genetic phenomena may be justified. Likewise, in repeated bioassays with microbial products, consistent shifts in responses at the LD_{50} must be observed before concerns about product quality would be justified (see Chapter 8).

Robertson et al.[2] showed six sets of results that demonstrate natural variation in diamondback moth (*Plutella xylostella* [L.], exposed to *Bacillus thuringiensis* subsp. *kurstaki*); Colorado potato beetle (*Leptinotarsa decemlineata* [Say], exposed to *B. thuringiensis* subsp. *tenebrioni*); and western spruce budworm (treated with pyrethrins, mexacarbate, or dichlorodiphenyltrichloroethane [DDT]). For diamond-back moth (late second to early third instar), the means (95% limits of natural varia-tion) of LD_{50}s based on these results were 0.36 (0.04–0.67) µg/g of diet; for second instar Colorado potato beetle, values were 80.0 (0–153) µg/g of diet. For western spruce budworm (sixth instar), results of bioassays with pyrethrins, mexacarbate, or DDT indicate that the means (95% limits of natural variation) of LD_{50}s were 0.07 (0.03–0.10), 0.11 (0.07–0.15), and 2.35 (0.6.9) µg/µl of solvent, respectively. The mean (95% limits of natural variation) at residual time $(RT)_{50}$ for diamondback moth was 2.0 (0.97–3.1); for Colorado potato beetle, limits were much wider at 3.97 (0.32–7.62). When pyrethrins were topically applied to western spruce budworm, limits of natural variation at RT_{50} were quite narrow—1.81 (0.84–2.78). Limits for mexacar-bate were intermediate—1.63 (0.85–2.41) and very wide for DDT—2.1 (≈0–4.24).

6.3.3 Developmental Stage

Biotic and abiotic factors affect arthropod development, with the result that a range of different developmental stages is present at any given time. A primary source of variation in response at the population level is differential responses of these life stages. Just after the Godfather laboratory colony was safely established, Paula had Jessica mix a conventional insecticide, permethrin, as a commercial for-mulation (MadMax SC) into artificial diet upon which larvae fed for the next seven days. The results of their bioassays (see Table 6.2) were typical of the results usually observed by many other investigators (see, e.g., Robertson and Boelter,[5] Elek and Beveridge,[6] and Biddinger et al.[7]): younger stages were most susceptible and the last instar was the least susceptible. This source of variation is one of the reasons that Paula, Jessica, and other investigators (see, e.g., Bolin et al.[8] and Toledo et al.[9]) use only one instar or stage in their experiments to identify possible significant differ-ences in arthropod response related to host plant varieties, production lots of micro-bial pesticides, and source or genetic status of different populations.

Life table analyses have been adapted to identify causes of mortality at a specific age interval (see, e.g., Morris and Miller[4]), and demographic toxicology has devel-oped as an approach to define the effects of toxicants at the population level (see Stark and Banks[10]). One way to understand the differences between quantal response

Table 6.2 Results of Bioassays with MadMax Incorporated into Diet upon which Godfather Larvae Fed

Instar	n	Slope (SE)	LD_{50} (95% CL)
1	500	2.23 (0.24)	0.0084 (0.0056–0.014)
2	625	3.65 (0.39)	0.061 (0.054–0.067)
3	376	2.03 (0.24)	0.14 (0.091–1.3)
4	425	2.19 (0.23)	0.19 (0.051–0.34)
5	525	2.37 (0.35)	0.22 (0.18–0.26)
6	400	1.89 (0.27)	0.40 (0.21–0.65)

bioassays and the models used in demographic toxicology is to compare their purposes. Quantal response bioassays are done to identify the significance of variables within the framework of statistical models. In contrast, demographic toxicology uses conceptual ecological models to predict effects at the population level. These two approaches need not be mutually exclusive.

Some bioassay techniques (e.g., topical application, diet incorporation) produce data that are not amenable for use in forecasts of the effects of age structure on population response. Other methods, such as spray application to insects on foliage, can produce data that are a better simulation of conditions in the field (e.g., Robertson and Rappaport[11]). Results of foliage-spray bioassays (Robertson et al.[12]) were used to develop a contour plot method to show the variability of response within a population based on both time and age structure. This method is as follows: For a given pair (x, y) of days x after emergence and concentration y, the probability $P(D \mid x, y)$ of an insect dying (D) before reaching adulthood was calculated as

$$P(D \mid x, y) = \sum_{i=2}^{6} P(D \mid I_i, y) P(I_i \mid x),$$

where $P(D \mid I_i, y)$ is the probability that an insect in instar i, sprayed with a concentration y, died before successful adult emergence and $P(I_i \mid x)$ is the probability that the insect is in instar i x days after emergence. $P(D \mid I_i, y)$ is estimated by $\hat{P}(D \mid I_i, y) = \hat{C} + (1 - \hat{C})\Phi(\hat{\alpha}_i + \hat{\beta}_i \log y)$—that is, the probit concentration–response curve for the ith instar, where C is mortality observed in controls and $P(I_i \mid x)$ is the relative frequency of instar i on day x.

6.4 EFFECTS OF NATURAL VARIATION ON PRODUCT QUALITY

Large-scale commercial production of any material used for arthropod population management involves strict measures to ensure product quality. For the biological insecticide industry, the most reliable method to estimate product quality during production is arthropod bioassay because no chemically based analytical methods are available to determine biological activity. According to usual practice, an aliquot from each lot is tested in four replicated bioassays and average potency of the lot is then computed.[8] That single number, the average, is the only potency reported for the lot.

A practical guideline that also more than exceeds the sample size recommended for screening experiments outlined in Chapter 5 (see Section 5.1.2) is to use at least six doses per bioassay, each with 30 test subjects per dose, with at least four replications. The total sample size of such a bioassay as a whole is 720. As long as each bioassay is done with separate dilutions from the initial aliquot, the experiment can be considered to be replicated regardless of whether two bioassays are done on the same day.

Assuming that the samples in the bioassays for the lot represent a random sample of responses of n present (and future) bioassays of the same lot, the approximate 95% limits of natural variation are mean ± 1.96 (SD), where mean and SD are the mean and standard deviation of the LDs from n samples. When LD ratios are used, as in bioassays with microbial preparations, the equation is the same, with mean and SD calculated from the lethal ratios from the n samples. For example, suppose that four replicated bioassays with lot #707 of *B.t. giganticus* produce LD_{50}s of 12,250, 11,671, 10,600, and 14,310 IUs. The mean and SD are 12,207 and 1559, with an approximate 95% CL of natural variation of 9151 to 13,766 IUs. When the label for the product is finalized, mean potency and the 95% interval should be reported so that variation is clearly acknowledged to exist. However, formulation instructions of the commercial product will continue to be based on its mean potency.

REFERENCES

1. Finney, D. J., *Statistical Methods in Biological Assay*, Griffin, London, 1964, pp. 91–92.
2. Robertson, J. L., Preisler, H. K., Ng, S. S., Hickle, L. A., and Gelernter, W. D., Natural variation: A complicating factor in bioassays with chemical and microbial pesticides, *J. Econ. Entomol.* 88, 1, 1995.
3. Price, P. W., *Insect Ecology*, John Wiley and Sons, New York, 1997.
4. Morris, R. F. and Miller, C. A., The development of life tables for the spruce budworm, *Can. J. Zool.* 32, 283, 1954.
5 Robertson, J. L. and Boelter, L. M., Toxicity of insecticides to Douglas-fir tussock moth, *Orgyia pseudotsugata* (Lepidoptera: Lymantriidae) I. Contact and feeding toxicity, *Can. Entomol.* 111, 1145, 1979.
6. Elek, J. and Beveridge, N., Effect of *Bacillus thuringiensis* subsp. *tenebrioni* insecticidal spray on the mortality, feeding, and developmental rates of larval Tasmanian eucalyptus leaf beetles (Coleoptera: Chrysomelidae), *J. Econ. Entomol.* 92, 1062, 1999.
7. Biddinger, D. J., Hull, L. A., and Rajotte, E. G., Stage specificity of various insecticides to tufted apple bud moth larvae (Lepidoptera: Tortricidae), *J. Econ. Entomol.* 91, 200, 1998.
8. Bolin, P. C., Hutchison, W. D., and Andow, D. A., Long-term selection for resistance to *Bacillus thuringiensis* Cry1Ac endotoxin in a Minnesota population of European corn borer (Lepidoptera: Crambidae), *J. Econ. Entomol.* 92, 1021, 1999.
9. Toledo, J., Liedo, P., Williams, T., and Ibarra, J., Toxicity of *Bacillus thuringiensis* δ-exotoxin to three species of fruit flies (Diptera: Tephritidae), *J. Econ. Entomol.* 92, 1052, 1999.
10. Stark, J. D. and Banks, J. E., Population-level effects of pesticides and other toxicants on arthropods, *Ann. Rev. Entomol.* 48, 505, 2003.
11. Robertson, J. L. and Rappaport, N. G., Direct, indirect, and residual toxicities of insecticide sprays to western spruce budworm, *Choristoneura occidentalis* (Lepidoptera: Tortricidae), *Can. Entomol.* 111, 1219, 1979.
12. Robertson, J. L., Richmond, C. E., and Preisler, H. K., Lethal and sublethal effects of avermectin B1 on the western spruce budworm (Lepidoptera: Tortricidae), *J. Econ. Entomol.* 78, 1129, 1985.

CHAPTER 7

Invasive Species Statistics

On a farm just outside Schaefferville, crop consultant Bill Emerine checks his pheromone traps on a weekly basis for common row crop pests—corn earworm, black cutworm, and European corn borer. This year, he has been asked by the state entomologist at the Missouri Department of Agriculture to scout for Silver Y moth (*Autographa gamma* [L.]), a highly polyphagous defoliator of many cultivated plants. Its accidental introduction to Missouri may pose concern in particular to soybean and cotton. There are no records of establishment in the United States; however, this species has been intercepted hundreds of times at the US ports of entry on imported vegetables such as garden pea (*Pisum sativum*), sugar beet (*Beta vulgaris*), and cabbage (*Brassica oleracea*). They can reduce crop yields by damaging leaves and are often considered to be a pest.

Three weeks into the growing season, several moths are caught in the Silver Y moth traps in his "Bad Boy" soybean variety fields. Mr. Emerine brings the samples to Paula for identification. The good news is that they are not Silver Y moth. The bad news is that they are another invasive species, *Rachiplusia ou* (gray looper), that most likely migrated from Kentucky. Larvae very closely resemble soybean loopers (*Chrysodeixis includens),* and each can devour a soybean leaf in 24 hours.

Invasive species problems are a new subject for Dr. Maven. She has not considered the bioassays that must be done to ensure that gray looper moth larvae do not become established in Missouri, nor is she familiar with US Department of Agriculture (USDA) quarantine or pest exclusion regulations. A short literature review[1,2] results in her realizing that quarantine of a particular species is not strictly a matter of its life and death after treatment. Instead, as indicated by Worner,[3] decisions are based on biological, economic, and environmental features of a region or country. Invasive species statistics concern not only the relationships between dose and response or time and response (see Chapter 11) but population ecology as well. However, an antiquated and overly stringent requirement that was imposed over 50 years ago for fruit heavily infested with one species of Tephritid fruit fly has stifled research in invasive species and quarantine statistics until relatively recently.

7.1 PROBIT OF 9

The "probit of 9" was codified by the USDA as the requirement for "probit 9 security." In percentages, this probit represents a mortality of 99.9968, or a survival of approximately 32 of 1,000,000, which would ensure that accidental introduction of an exotic species of arthropod or plant disease would not be possible. The same requirement still remains as the legal basis of USDA plant protection and quarantine treatment schedules.[4] Three related assumptions are inherent in the probit 9 requirement: (1) the probit model is always appropriate for data analyses from commodity treatment bioassays; (2) probability of death is the only criterion relevant to the future establishment of a pest species in a new environment; and (3) 99.9968% effectiveness is the minimum level necessary for quarantine security.

Even if the first two assumptions were true, no investigations have ever shown that the probit model is a matter of universal truth. As we indicated in Chapter 3, the debate over probit versus logit analysis has been ongoing among statisticians for over 50 years, and other alternatives have been proposed. Robertson et al.[5,6] have suggested that a general term, Q, be substituted for the word *probit* in the context of insect quarantine. This simple change in notation would eliminate the implication that the standard normal is the only distribution function that can or should be used to estimate 99.9968% mortality. We will refer to the lethal dose, $LD_{99.9968}$, as the Q_9 here.

7.1.1 Laboratory Bioassays to Estimate Q_9 in a Confirmatory Test

The progression of testing for invasive species problems is the same for any other treatments in pest management. Laboratory tests are used to test possible treatments,[7] identify the most susceptible life stages of the arthropod[8] that is to be excluded from a new area, and, if the treatment appears to be sufficiently effective, recommend the dose to be tested in a larger test to confirm treatment efficacy.

Until relatively recently, different varieties of a commodity were considered as different entities with respect to a treatment; a different test protocol was necessary for each variety. Each was subjected to a separate (unreplicated) confirmatory test. Evidence presented by Tebbetts et al.,[9,10] Yokoyama et al.,[11] Maindonald et al.,[12] and Robertson and Yokoyama[13] has shown conclusively that varietal differences are no more than a reflection of natural variation in response of the insect. Periodically, an importing nation will attempt to impose trade barriers based on results of bioassays of an arthropod on a variety of a particular commodity. In 1997–98, such an issue was finally resolved with the government of Japan through negotiations at the World Trade Organization.[14] Paula Maven must develop the dose–response testing to be used to predict the treatment intensity for large-scale testing. Whatever the protocol, Jessica may have to do a quarantine treatment efficacy bioassay to meet federal requirements.

7.1.1.1 Differences in Estimates Depending on Tolerance Distribution

Robertson et al.[5,6] used data from three typical bioassays to show differences at the Q_9 when the probit or logit model was used (see Table 7.1). The bioassays were two

Table 7.1 Results for Seven Generations of *B. dorsalis* Treated with Cold

Generation	Q_9 (95% CL)	χ^2	df
30	9.52 (8.79–10.24)	2.68	6
31	10.64 (8.47–12.82)	1.52	4
32	9.46 (8.71–10.21)	8.96	7
33	9.38 (8.46–10.30)	2.54	5
34	9.30 (7.89–10.70)	1.69	6
35	10.10 (9.17–11.04)	3.12	6
37	10.54 (10.34–10.75)	7.25	9

Source: From Robertson JL, Preisler HK, Frampton ER, Armstrong JW, Statistical analyses to estimate efficacy of disinfestation treatments, in Sharp JL, Hallman GJ, eds., *Quarantine Treatments for Pests of Food Plants*, Westview Press, Boulder, CO, 1994, 47–65.[6]

Note: Estimates were obtained by fitting a probit model with a random effect factor to account for extra-binomial variation.

commodity treatment experiments: (1) *Bactrocera dorsalis* (Hendel), exposed to heat treatment for increasing times in days, and *Ceratitis capitata* (Wiedemann) in mangoes immersed in hot water for increasing times in minutes; and (2) topical application of a pesticide (mexacarbate) on western spruce budworm. The choice of a model had a large effect on estimates of Q_9, especially in the bioassay with western spruce budworm.

For the two generations of *B. dorsalis*, the Q_9 estimates were substantial despite use of ≥28,000 individuals in each bioassay and regardless of model assumed. Large values of the χ^2 goodness-of-fit statistic versus degrees of freedom indicated that neither the probit nor logit model adequately fit the data. Such large χ^2 values are usually managed by multiplication of variances and covariances by the heterogeneity factor (SAS[15]; Sharp and Picho-Martinez[16]). This practice assumes that the causes of poor fit affect only the calculations of the standard error, but not each point estimate. Residual plots for both the F_{30} and F_{31} generations showed evidence of extrabinomial variation, so further analyses of seven generations was done with a compound probit model that incorporates a random error effect at the level of dose (see Preisler[17]). Even when the random effects model was used, estimates of Q_9 differed (see Table 7.1). This observation is consistent with the general phenomenon of natural variation in response to chemical and microbial agents reported by Robertson et al.[18]

7.1.1.2 General Formula for Selection of Dose in a Confirmatory Test Based on Laboratory Bioassays

For selection of dose (or time at a given dose) to be tested in the next phase (confirmatory tests), Robertson et al.[5] suggested that computation of the mean value of Q_9 and its SD could be used along with the value of $x\pi$. Assuming that the samples included in the experiment provide a random sample of the responses of present and future generations of n insect populations, the confirmatory dose is $(Q_9)/n + 1.96(SD)$, where $x\pi$ is the πth percentile point of the distribution curve. Thus, for the seven *B. dorsalis* generations, 95% of the generations would have a Q_9 that is less than $9.855 + (1.65 \times 0.56) = 10.78$ days. The confirmatory test to verify that quarantine security requirements are met would be done with 2.8°C for a period of 10.78 days.

7.1.1.3 Dose Placement and Sample Size Requirements for Estimation of Q_9 in Bioassays

Dr. Maven decides to use OptiDose to examine the problems of dose placement and sample size that she would ask Jessica to test if their grant application is funded. Because soybeans are an important cash crop for Missouri, Paula and Jessica start to design their protocols.

Based on the general information about specialized bioassays to estimate the LD_{99}, Paula uses a seven-dose design with cell sizes from 1500 to 4000 (total sample sizes, 10,500–28,000) to examine where doses should be placed based on the width of L^* (see Table 7.2). These results show that all of the designs except the one that includes 0.80 as the lowest dose provide acceptable precision for total sample sizes from 10,500 to 28,000. However, the highest dose in each design must have five survivors out of 10,000 individuals. For precision, each bioassay should include 70,000 individuals. Realizing that the research center will not build another $100,000 wing onto their building just to house more moths, Paula considers testing a smaller number of individual larvae.

Smaller sample sizes can be tested with the understanding that no extrapolation is permissible and that doses should be placed approximately as shown in Table 7.2. However, even the use of 3000 individuals (such as would be needed for precise estimation of the LD_{99}; see Section 5.3) would be beyond the capacity of most laboratory rearing programs. Even if Jessica wanted to test a treatment with gray looper moth larvae, she could not do so simply because of the sample sizes required. Paula concludes that the arbitrary standard that was set for one species of Tephritid fruit fly is not universally applicable to other insect species, particularly Lepidoptera.

Suggestions for alternatives to treatment verification by killing large numbers of individuals have included the likelihood of mating pairs of insects surviving in a single consignment,[19] low pest prevalence on poor hosts,[20] modeling numbers of pests arriving in susceptible areas,[21] and "maximum pest limit."[22] When treatment verification to appropriate numbers may not be possible, the analysis of carefully designed dose–response experiments may be used to define appropriate treatment dosages (see Section 5.4).

7.1.1.4 Varietal Differences

Natural variation in response occurs whenever a bioassay is repeated (see Chapter 6).[13] Evidence that cultivars significantly affect dose–responses of an arthropod subjected to quarantine treatments is overwhelming.[9–12] Based on the procedures described by Robertson and Yokoyama,[13] Paula gives Jessica the bioassay method to use. She will test key insecticides in diet-incorporated bioassays of gray looper

Table 7.2 Confidence Interval Estimates for Q_9 in a Seven-Dose Design

Cell	Sample	P_0	L^* Probit
1500	10,500	0.0125, 0.90, 0.925, 0.95, 0.97, 0.99, 0.9995	0.263
		0.025, 0.90, 0.925, 0.95, 0.97, 0.99, 0.9995	0.241
		0.05, 0.90, 0.925, 0.95, 0.97, 0.99, 0.9995	0.230
		0.10, 0.90, 0.925, 0.95, 0.97, 0.99, 0.9995	0.228
		0.25, 0.90, 0.925, 0.95, 0.97, 0.99, 0.9995	0.245
		0.50, 0.90, 0.925, 0.95, 0.97, 0.99, 0.9995	0.302
		0.80, 0.90, 0.925, 0.95, 0.97, 0.99, 0.9995	0.513
2000	14,000	0.0125, 0.90, 0.925, 0.95, 0.97, 0.99, 0.9995	0.228
		0.025, 0.90, 0.925, 0.95, 0.97, 0.99, 0.9995	0.209
		0.05, 0.90, 0.925, 0.95, 0.97, 0.99, 0.9995	0.199
		0.10, 0.90, 0.925, 0.95, 0.97, 0.99, 0.9995	0.197
		0.25, 0.90, 0.925, 0.95, 0.97, 0.99, 0.9995	0.212
		0.50, 0.90, 0.925, 0.95, 0.97, 0.99, 0.9995	0.261
		0.80, 0.90, 0.925, 0.95, 0.97, 0.99, 0.9995	0.444
3000	21,000	0.0125, 0.90, 0.925, 0.95, 0.97, 0.99, 0.9995	0.186
		0.025, 0.90, 0.925, 0.95, 0.97, 0.99, 0.9995	0.171
		0.05, 0.90, 0.925, 0.95, 0.97, 0.99, 0.9995	0.162
		0.10, 0.90, 0.925, 0.95, 0.97, 0.99, 0.9995	0.162
		0.25, 0.90, 0.925, 0.95, 0.97, 0.99, 0.9995	0.173
		0.50, 0.90, 0.925, 0.95, 0.97, 0.99, 0.9995	0.213
		0.80, 0.90, 0.925, 0.95, 0.97, 0.99, 0.9995	0.444
3500	24,500	0.0125, 0.90, 0.925, 0.95, 0.97, 0.99, 0.9995	0.172
		0.025, 0.90, 0.925, 0.95, 0.97, 0.99, 0.9995	0.158
		0.05, 0.90, 0.925, 0.95, 0.97, 0.99, 0.9995	0.150
		0.10, 0.90, 0.925, 0.95, 0.97, 0.99, 0.9995	0.149
		0.25, 0.90, 0.925, 0.95, 0.97, 0.99, 0.9995	0.161
		0.50, 0.90, 0.925, 0.95, 0.97, 0.99, 0.9995	0.197
		0.80, 0.90, 0.925, 0.95, 0.97, 0.99, 0.9995	0.334
3750	26,250	0.0125, 0.90, 0.925, 0.95, 0.97, 0.99, 0.9995	0.166
		0.025, 0.90, 0.925, 0.95, 0.97, 0.99, 0.9995	0.153
		0.05, 0.90, 0.925, 0.95, 0.97, 0.99, 0.9995	0.145
		0.10, 0.90, 0.925, 0.95, 0.97, 0.99, 0.9995	0.144
		0.25, 0.90, 0.925, 0.95, 0.97, 0.99, 0.9995	0.155
		0.50, 0.90, 0.925, 0.95, 0.97, 0.99, 0.9995	0.191
		0.80, 0.90, 0.925, 0.95, 0.97, 0.99, 0.9995	0.323
4000	28,000	0.0125, 0.90, 0.925, 0.95, 0.97, 0.99, 0.9995	0.161
		0.025, 0.90, 0.925, 0.95, 0.97, 0.99, 0.9995	0.148
		0.05, 0.90, 0.925, 0.95, 0.97, 0.99, 0.9995	0.141
		0.10, 0.90, 0.925, 0.95, 0.97, 0.99, 0.9995	0.139
		0.25, 0.90, 0.925, 0.95, 0.97, 0.99, 0.9995	0.150
		0.50, 0.90, 0.925, 0.95, 0.97, 0.99, 0.9995	0.185
		0.80, 0.90, 0.925, 0.95, 0.97, 0.99, 0.9995	0.313

Table 7.3   Results of Soybean Cultivar Tests with Gray Looper in CATT Tests

Cultivar	n_c	N	Slope (SE)	LD$_{50}$ (95% CL) (Time in Minutes)	LD$_{50}$ Ratio
BadBoy	97	495	3.180 (0.423)	9.4 (6.6–11.8)	
Hermione	92	422	2.543 (0.308)	11.5 (7.7–15.1)	0.846 (0.697–1.027)
BadBoy	94	492	3.622 (0.500)	11.3 (8.8–13.7)	
Joli	94	561	3.093 (0.468)	23.7 (11.5–30.7)	0.477 (0.382–0.594)
BadBoy	90	498	2.155 (0.186)	16.8 (10.6–23.7)	
Milo	90	596	1.998 (0.202)	4.3 (2.7–5.8)	3.193 (2.942–5.204)
BadBoy	95	429	2.463 (0.312)	9.6 (3.7–14.3)	
Yumi	94	566	3.142 (0.318)	10.0 (8.0–11.9)	0.958 (0.772–1.190)

moth larvae on five soybean varieties ("BadBoy," "Joli," "Hermione," "Yumi," and "Milo"). The objective of a test of varietal differences is comparison and nothing more, so the best point of comparison is *not* Q_9 but LD$_{50}$.

The general design of this type of experiment should include pairwise comparisons of gray looper larvae reared on each cultivar versus those reared on BadBoy so that the natural variation of BadBoys can be estimated, together with an overall comparison of relative susceptibility of gray looper larvae that feed on the other cultivars. If enough gray looper larvae were available (and if they fed on BadBoys), Jessica would test about 500 individuals in the same way that she does a screening experiment with a chemical. Hypothetical results of such analyses with the PoloJR software program are shown in Table 7.3.

Hermione and Yumi have LD$_{50}$s that are equal to those of BadBoys in their respective experiments. The response of Joli is significantly longer than that of BadBoys, and that of Milos is significantly shorter. However, the mean ± SD of the LD$_{50}$s of BadBoys in these experiments is 11.7 ± 1.7 days. Using the general equation for natural variation discussed in Chapter 6, and assuming that the cultivars represent a random sample of past and future cultivars upon which BadBoys might feed, natural variation would be reflected by any LD$_{50}$ between 8.4 and 15.3 days.

7.1.2 Confirmatory Tests

In a confirmatory test, large numbers of the target species are tested with a fixed treatment level (usually the Q_9), and the upper 95% boundary for the probability of at least one survivor is calculated. For example, if n individuals are treated and no survivors are observed, the 95% upper boundary is $1 - (0.05)^{1/n}$. If $n = 95,000$, for example, then $p_u = 31.53$ per million. To meet the Q_9 criterion, <32 survivors per million are required. Any assumptions about a particular tolerance distribution are unnecessary at the stage of confirmatory testing.

The correct equation[20] to relate sample size (n) and the number of survivors (s) observed in confirmatory tests to true survivorship (p_u) where one or more survivors is observed is

$$\sum_{x=0}^{x=s} e^{-m} m^x / x! = 1 - C.$$

(7.1)

In this equation, m is the product of n and p_u at a given confidence limit C (usually 0.95) for s survivors. The methods of Couey and Chew[23] can be used to calculate p_u.

7.2 ECOLOGICAL APPROACHES TO INVASIVE SPECIES: RISK

Death is not the only criterion for the effectiveness of quarantine treatments. Scientific knowledge about exotic pest species increased substantially between 1939[24–26] and the present. These ecological studies clearly showed that mortality is not the only criterion that should be used to evaluate treatment efficacy.[24–29] As Paula Maven is well aware, the science of ecology is relatively new.[24,25] With a more ecological approach, emphasis in applied entomology shifted from control of arthropod pests to arthropod population management in the mid-1970s. In invasive species entomology, the role of ecological parameters such as percentage infestation of a commodity, host suitability, pest distribution in a commodity, and arthropod survival on various hosts indicated that a distinction must be made between commodities that are good hosts or that are heavily infested, versus those that are poor hosts or that are rarely infested.

7.2.1 The Alternative Efficacy Approach for Species on Poor Hosts

The alternative efficacy approach measures risk as the probability of a mating pair, gravid female,[26,30] or parthenogenic individual surviving in a shipment. Many factors can be involved in calculations of risk. These include lot size, mean numbers of arthropods per unit of the commodity, and the proportion of the commodity infested.[27,29] One general expression of the risk of infestation is

$$\frac{e^{-\lambda} \lambda^r}{r!},$$

(7.2)

where $\lambda = N\varphi\mu p$, N = lot size, φ = treatment efficacy, μ = mean number of arthropods per unit of the commodity, and p = proportion of the units infested. Therefore,

$$q = \Pr[r \geq 2] - \left(1 - e^{-\lambda/2}\right)^2.$$

(7.3)

A modification of this equation and the subsequent calculations is shown in Equations 1 through 4 of Follett and McQuate[30]; these equations can then be used to calculate pest risk and the numbers of insects that must be tested to reduce the numbers of test subjects required for quarantine treatment development when the commodity is a poor host for the arthropod. Such arthropod pest–host combinations include *B. dorsalis* (Hendel) on rambutan (*Nephelium lappacium* L.),[30] oriental fruit fly on avocados,[31] and codling moth on nectarines[32] or sweet cherries.[33]

7.2.2 Risk for Lethal and Sublethal Effects on Beneficial Insects

Natural enemies of invasive species become collateral damage when the toxic effects of insecticides are not quantified for their susceptibility. Parasitoids (*Trichogramma* sp.) are an especially important natural enemy of many lepidopteran pests throughout the world. Risk quotient analysis [i.e., recommended field rate $(g/AI/ha)/LC_{50}$ of beneficial insect (mg AI L^{-1})] has been used to determine the toxicity of select insecticides on the parasitoid *Trichogramma confusum* (Viggiani).[34] Health and mobility of predator species can by quantified by the combination of dose response, time, and observation expectations.[35]

7.2.3 Risk for Preferred Hosts and Heavy
Infestations: Systems Approach

Mangan et al.[36] describe a landmark investigation in which Equations 7.1 through 7.3 were used to show that probability of accidental introduction of a single mating pair of Mexican fruit fly (*Anastrepha ludens* [Loew]) was so high that insecticide application that produced a Q_9, sterile insect release, and selective harvest of fruit combined would allow the survival of fewer than one reproductive pair of flies per shipment of mangoes and citrus. This study is unique because it provides the only example of a true scientific systems approach, including risk analysis, applied to an actual commercial cropping system. Robertson et al.[5,6] had recommended that such analyses were crucial to the improvement of scientific pest exclusion systems in entomology.

7.3 CONCLUSIONS

The probit 9 and the automatic requirement for 99.9968% mortality in a quarantine treatment are two of the last lingering remnants of the antiquated pest management practices of the early twentieth century. As a USDA staff entomologist, A. C. Baker was doubtless familiar with the use of Paris green and other arsenical pesticides on crops and export commodities. Once the USDA had a regulation in place, dichlorodiphenyltricholoroethane and later synthetic organic pesticides replaced arsenicals as the control agents of choice on crops and as fumigation agents. Rather than a practice based on sound science, the probit 9 requirement is an anachronism based in bureaucracy.

If invasive species entomology had developed to provide knowledge to producers and potential exporters (or importers) in the same way that epidemiology has developed in conjunction with medical science, the statistics of risk analysis would have been the primary methods described here. As this entomological specialty actually has evolved, risk analyses have been a recent addition. The studies of Landolt et al.,[26]

Vail et al.,[27] and Baker et al.[28] were just the initial steps toward building the scientific basis that will be the postmortem for the probit 9. Even though Mangan et al.[36] assumed that 99.9968 mortality had occurred in their study, the statistical methods that they used will allow the development of even more thorough risk analyses in the future.

Paula Maven concludes that the three assumptions inherent in the probit 9 are unfounded. The probit model is not always appropriate for data analyses from commodity treatment bioassays because (1) the normal distribution is one of many available, and there is no basis to assume that (2) probability of death is the only criterion relevant to the future establishment of a pest species in a new environment; and (3) 99.9968% effectiveness is the minimum level necessary for pest exclusion.

REFERENCES

1. Paull, R. E. and Armstrong, J. W., eds., *Insect Pests and Horticultural Products: Treatments and Responses*, CAB International, Wallingford, England, 1994.
2. Sharp, J. L. and Hallman, G. J., eds., *Quarantine Treatments for Pests of Food Plants*, Westview Press, Boulder, CO, 1994.
3. Worner, S. P., Predicting the establishment of exotic pests in relation to climate, in Sharp, J. L. and Hallman, G. J., eds., *Quarantine Treatments for Pests of Food Plants*, Westview Press, Boulder, CO, 1994, pp. 11–31.
4. USDA (United States Department of Agriculture), *Plant Protection and Quarantine Treatment Manual*, USDA, APHIS, Washington, DC, 1998.
5. Robertson, J. L., Preisler, H. K., and Frampton, E. R., Statistical concept and minimum threshold concept, in Paull, R. E. and Armstrong, J. W., eds., *Insect Pests and Horticultural Products: Treatments and Responses*, CAB International, Wallingford, England, 1994, pp. 47–65.
6. Robertson, J. L., Preisler, H. K., Frampton, E. R., and Armstrong, J. W., Statistical analyses to estimate efficacy of disinfestation treatments, in Sharp, J. L. and Hallman, G. J., eds., *Quarantine Treatments for Pests of Food Plants, Westview Press,* Boulder, CO, 1994, pp. 47–65.
7. Zettler, J. L., Follett, P. A., and Gill, R. F., Susceptibility of *Maconellicoccus hirsutus* (Homoptera: Pseudococcidae) to methyl bromide, *J. Econ. Entomol.* 95, 1169, 2002.
8. Neven, L. G., Combined heat and controlled atmosphere quarantine treatments for control of codling moth in sweet cherries, *J. Econ. Entomol.* 98, 709, 2002.
9. Tebbets, J. S., Hartsell, P. L., Nelson, H. D., and Tebbets, J. C., Methyl bromide fumigation of tree fruits for control of Mediterranean fruit fly, *J. Agric. Food Chem.* 31, 247, 1983.
10. Tebbets, J. S., Vail, P. V., Hartsell, P. L., and Nelson, H. D., Dose/response of codling moth (Lepidoptera: Tortricidae) eggs and nondiapausing and diapausing larvae to fumigation with methyl bromide, *J. Econ. Entomol.* 79, 1039, 1986.
11. Yokoyama, V. Y., Miller, G. T., and Hartsell, P. L., Evaluation of a methyl bromide quarantine treatment to control codling moth (Lepidoptera: Tortricidae) on nectarine cultivars proposed for export to Japan, *J. Econ. Entomol.* 83, 466, 1990.

12. Maindonald, J. H., Waddell, B. C., and Birtles, D. B., Response to methyl bromide fumigation of codling moth (Lepidoptera: Tortricidae) eggs on cherries, *J. Econ. Entomol.* 85, 1222, 1992.

13. Robertson, J. L. and Yokoyama, V. Y., Comparison of methyl bromide LD_{50}s of codling moth (Lepidoptera: Tortricidae) on nectarine cultivars as related to natural variation, *J. Econ. Entomol.* 91, 1433, 1998.

14. SICE Dispute Settlement WTO Report, Japan—Measures affecting agricultural products, report of the panel, *Probit 9, Dose–Mortality Tests and Confirmatory Tests*, http://www.sice.oas.org/dispute/wto/ds76/76r07.asp, 1999.

15. SAS Institute, http://www.sas.com/technologies/analytics/statistics/stat.

16. Sharp, J. L. and Picho-Martinez, H., Hot-water quarantine treatment to control fruit flies in mangoes imported into the United States from Peru, *J. Econ. Entomol.* 83, 1940, 1990.

17. Preisler, H. K., Assessing insecticide bioassay data with extra-binomial variation, *J. Econ. Entomol.* 81, 759, 1988.

18. Robertson, J. L., Preisler, H. K., Ng, S. S., Hickle, L. E., and Gelernter, W. D., Natural variation: A complicating factor in bioassays with chemical and microbial pesticides, *J. Econ. Entomol.* 88, 1, 1995.

19. Landolt, P. J., Chambers, D. L., and V. Chew., Alternative to the use of probit-9 mortality as a criterion for quarantine treatments of fruit fly (Diptera: Tephritidae)-infested fruit, *J. Econ. Entomol.* 77, 285–287, 1984.

20. Rollett, P. A. and McQuate, G. T., Accelerated development of quarantine treatments for insects on poor hosts, *J. Econ. Entomol.* 94, 1005–1011, 2001.

21. Hennessey, M. K., Quarantine pathway pest risk analysis at the APHIS Plant Epidemiology and Risk Analysis Laboratory, *Weed Technol.* 18, 1484–1485, 2004.

22. Baker, R. J., Cowley, J. M., Harte, D. W., and Frampton, E. R., Development of a maximum pest limit for fruit flies (Diptera: Tephritidae) in produce imported into New Zealand, *J. Econ. Entomol.* 83, 13–17, 1990.

23. Couey, H. M. and Chew, V., Confidence limits and sample size in quarantine research, *J. Econ. Entomol.* 79, 887, 1986.

24. Andrewartha, H. G. and Birch, L. C., *The Distribution and Abundance of Animals*, University of Chicago Press, Chicago, 1954.

25. Huffacker, C. B. and Rabb, R. L., eds., *Ecological Entomology*, John Wiley & Sons, New York, 1984.

26. Landolt, P. J., Chambers, D. L., and Chew, V., Alternatives to the use of probit 9 mortality as a criterion for quarantine treatments of fruit fly (Diptera: Tephritidae)–infested fruit, *J. Econ. Entomol.* 77, 285, 1984.

27. Vail, P. V., Tebbets, J. S., Mackey, B. E., and Curtis, C. E., Quarantine treatments: A biological approach to decision making for selected hosts of codling moth (Lepidoptera: Tortricidae), *J. Econ. Entomol.* 86, 70, 1993.

28. Baker, R. T., Cowly, J. M., Harte, D. S., and Frampton, E. R., Development of a maximum pest limit for fruit flies (Diptera: Tephritidae) in produce imported into New Zealand, *J. Econ. Entomol.* 83, 13, 1990.

29. Liquido, N. J., Barr, P. G., and Chew, V., CQT_STATS: Biological statistics for pest risk assessment in developing commodity quarantine treatment, *USDA-ARS Publ. Series* (available at http://www.pbarc.ars.usda.gov), 1996.

30. Follett, P. A. and McQuate, G. T., Accelerated development of quarantine treatments for insects on poor hosts, *J. Econ. Entomol.* 94, 1005, 2001.

31. Liquido, N. J., Chan, H. T., and McQuate, G. T., Hawaiian tephritid fruit flies (Diptera): Integrity of the infestation-free quarantine procedure for 'Sharwil' avocado, *J. Econ. Entomol.* 88, 85, 1995.
32. Curtis, C. E., Clark, J. D., and Tebbetts, J. S., Incidence of codling moth (Lepidoptera: Tortricidae) in packed nectarines, *J. Econ. Entomol.* 84, 1686, 1991.
33. Data and report cited by Follett, P. A. and McQuate, G. T., note 29 above.
34. Wang, Y., Chen, L., An, X., Jiang, J., Wang, Q., Cai, L., and Zhao, X., Susceptibility of selected insecticides and risk assessment in the insect egg parasitoid *Trichogramma confusum* (Hymenoptera: Trichogrammatidae), *J. Econ. Entomol.* 106, 142–149, 2013.
35. Eisenback, B. M., Salom, S. M., Kok, L. T., and Lagalante. A. F., Lethal and sublethal effects of imidacloprid on hemlock woolly adelgid (Hemiptera: Adelgidae) and two introduced predator species, *J. Econ. Entomol.* 103, 1222–1234, 2010.
36. Mangan, R. L., Frampton, E. R., Thomas, D. B., and Moreno, D. S., Application of the maximum pest limit concept to quarantine security standards for the Mexican fruit fly (Diptera: Tephritidae), *J. Econ. Entomol.* 90, 1433, 1997.

CHAPTER 8

Statistical Analyses of Data from Bioassays with Microbial Products

Every university town seems to have a genetic engineering laboratory, and Schaefferville is no exception. Dr. Hans Gruber, a first cousin of Tobias Snape, is the director and production manager of Gruber Genetically Engineered Products. Hans's company produces *Bacillus thuringiensis* varieties for insect management. His staff isolated the strain that affects the Godfather larvae, and Hans names it *B. thuringiensis* var. *giganticus*. The federal registration process[1] begins, and the Gruber company's product, Worm-B-Gone, is registered for use on tomatoes and the other crops that are demolished by infestations of Godfather larvae.

Hans Gruber becomes concerned about the quality of the Worm-B-Gone production process after random batches from the fermentation vats are tested in laboratory bioassays. Results of bioassays with Worm-B-Gone and *B. thuringiensis* var. *kurstaki* (i.e., the standard) by more than one independent laboratory show inconsistent results. Tobias suggests Hans contact Paula Maven to review the idiosyncrasies involved with bioassays of microbial products.

8.1 BIOLOGICAL UNITS AND STANDARDS

Microbial pesticides are unique because their active ingredient is a mixture of large proteins whose activity is stated in terms of standardized units that must be estimated biologically rather than defined chemically. This difference is recognized in the U.S. Environmental Protection Agency guideline, but it merits a thorough examination of all the statistical procedures that are involved. Dr. Maven, mindful of all that she has recently discovered about natural variation in response, begins her review by reading about the analytical chemistry of *B. thuringiensis*.[1]

When formulated for a bioassay, the active ingredient of the toxicant is known in terms of its chemical purity. Thus, Jessica has applied a measured volume containing a precise amount of toxicant in each bioassay with sibling groups, population samples, different generations, and other samples. When she tests microbial pesticides, in contrast, the amount of active material is based on arbitrary units assigned to standard preparations. The historical development of standard bioassay methods for various strains of *B. thuringiensis* has been reviewed by Beegle,[2] but a critical assessment of the statistical methods used for data analyses of bioassays with these products has never been done and is urgently needed. Commercial development and production of microbial products depend not only on standardized bioassay procedures but also on the use of valid statistical methods.

Early in the development of products based on *B. thuringiensis*, investigators recognized the lack of a relationship between spore count and biological activity of a preparation. As a result, an alternative approach was adopted so that a standard preparation (expressed in arbitrary units called International Potency Units, or IUs) was tested concurrently with each test preparation, and a lethal dose, LD_{50}, for each preparation was estimated as a ratio. The ratio was then multiplied by the IUs of the standard to obtain the putative potency. This procedure avoids an extremely complex computational problem because the binary response for each test preparation is not simply $y_t = \alpha + \beta x_t$. Instead, a more complicated regression line is needed.

8.2 A REVISED DEFINITION OF RELATIVE POTENCY

Dr. Maven reads, "Standards are used to determine relative toxicities for different preparations, correct day-to-day fluctuations, and to some extent, correct for different assay conditions and methodologies between different laboratories."[2] Here, she detects a semantic problem. In the dictionary, and certainly by logic, *potency* should be the same as *toxicity*. In the context of probit and logit regression, however, the terms are *not* synonymous because Finney's[3] statistical definition of potency is predicated on the assumption that response lines are parallel. Recall that parallelism means that the models for toxic action of the two stimuli are the same.

At the time Finney's equations were published, calculations were derived and problems were solved by hand; modern computer technology had not yet made possible the solution of likelihood ratio tests to test the hypotheses of parallelism and equality.[4] Likewise, the toxicity ratio test (see Section 3.2.1.3) had not yet been derived. Invertebrate pathologists therefore assumed that a potency estimation was not valid if slopes were significantly different, while at the same time they lacked a statistical method to test the hypothesis of parallelism. As a result, solutions to this problem such as those discussed by Beegle[2] involve assumptions about corrections for parallelism that are both improper and unnecessary.

The terms *relative toxicity* and *relative potency* can be used synonymously so long as no assumptions about parallelism are made. Invertebrate pathologists have two choices: discard the assumption of parallelism because it is a statistical relic, or provide biological evidence that parallelism is necessary for a simple point comparison at the 50% level. The logical choice is to discard the unnecessary assumption and stop the dangerous practice of discarding data for no good reason.

The correct calculation for relative potency involves a simple modification of the calculations previously given (in Section 3.2.1.2), so that upper and lower 95% confidence limits are estimated along with mean relative potency. By convention, relative potency is estimated at the 50% response level. For example, Jessica has completed five bioassays in which the potency of experimental fermentation batches of *B. thuringiensis* var. *giganticus* must be estimated in comparison with the standard *B. thuringiensis* var. *kurstaki*. Results produced by probit analysis done with PoloJR[5] provide an ideal means to calculate the relative toxicities with 95% limits in terms of IUs. If other software is used for the data analyses, lethal dose ratios of each lot compared with the standard can be computed by the methods given in Section 3.2.1.2.

Because it is based on the assumptions of the probit model, mean potency is an estimate, not a measurement. Therefore, sampling error and natural variation will contribute to the appearance of different potency values from the same sample in repeated bioassays at the same time and to different values over time.

8.3 EFFECTS OF NATURAL VARIATION ON PRODUCT QUALITY

Efforts to standardize the laboratory methods used in bioassays of microbial products have been extraordinary and are part of the overall attempt to eliminate variation in response to standard microbial preparations of *B. t. thuringiensis*, *B. t. kurstaki*, *B. t. israeliensis*, and *B. tentomocidus*. The variables under strict control in a given laboratory are usually method of exposure, insect developmental stage, incubation conditions, and duration of exposure. For example, the microbial preparation can be applied to the surface of a meridic diet or mixed with a molten diet; the numerous other methods are described by Beegle.[2] In a specific bioassay, only one developmental stage of the arthropod is used and individuals in that stage are, as much as is possible, the same age. Even when all of these experimental variables are controlled, variable results in response to standard preparations will be observed in the same laboratory under exactly the same treatment conditions. Weekly repeated bioassays of *B. t. tenebrioni* on early-second-instar Colorado potato beetle and *B. t. kurstaki* on late-second to early-third-instar diamondback moth show significant difference in response (Tables 8.1 and 8.2).[6] The same type of variation can be expected when results from bioassays in one laboratory are compared with results from another laboratory, when results for the same microbial preparation are tested after

Table 8.1   Toxicity of *B. thuringiensis* subsp. *tenebrioni* to Colorado Potato Beetle over 83 Weeks

Week	n	Slope ± SE	LD_{50} (95% CI)
5	989	1.87 ± 0.15	20.1 (15.4–24.8)
4	898	2.26 ± 0.16	23.4 (20.4–26.3)
28	992	1.67 ± 0.15	30.9 (23.0–38.4)
6	981	2.30 ± 0.15	31.6 (25.3–38.5)
3	1444	1.13 ± 0.10	35.3 (10.6–59.7)
8	980	2.12 ± 0.15	39.5 (34.0–45.5)

Source: Reprinted from Robertson JL, Preisler HK, Ng SS, Hickle LA, Gelernter WD, *J. Econ. Entomol.*, 88, 1, 1995.
Note: CI, confidence interval.

Table 8.2   Responses of Diamondback Moth to *B. thuringiensis* subsp. *kurstaki* over 37 Weeks

Week	n	Slope ± SE	LD_{50} (95% CI)
3	1642	2.33 ± 0.14	0.18 (0.14–0.22)
19	724	3.22 ± 0.27	0.19 (0.14–0.23)
15	414	3.68 ± 0.34	0.26 (0.16–0.37)
4	1741	2.42 ± 0.12	0.26 (0.13–0.39)
7	576	4.29 ± 0.39	0.28 (0.25–0.30)
22	567	5.55 ± 0.45	0.28 (0.19–0.36)

Source: Reprinted from Robertson JL, Preisler HK, Ng SS, Hickle LA, Gelernter WD, *J. Econ. Entomol.*, 88, 1, 1995.

Note: The regression for week 3 had the lowest LD_{50} and was used for the standard for comparison for all other LD_{50}s. CI, confidence interval.

various periods of storage, or when the responses of populations are compared. As will be shown in Chapter 9, numerical differences in toxicity ratios or relative potencies alone cannot be used to identify resistance without further biological evidence and identification of the extent of natural variation.

8.4 BIOASSAYS FOR NONTARGET ORGANISMS OR HOST ANIMALS

Microbial insecticides have been instrumental for control of insects of medical and veterinary importance. The best-known pathogens include *Bacillus popilliae* (Dutky), *B. cereus* (Frankland and Frankland), and *B. thuringiensis* Berliner.[6] The latter has been widely used for mosquito,[7] horn fly,[8] and stable fly[9] control.

The efficacy of microbial insecticides can be tested using methods similar to those used for testing the efficacy of synthetic insecticides. Toxins, varieties, and strains can be compared and contrasted for their effects on larval mortality or morphology. Estimated LC_{50}s for toxins produced in strains of *B. israelensis* and *B. kurstaki* determined the most virulent for larvae of oriental rat flea, *Xenopsylla cheopis* (Rothschild), were calculated based on diet-incorporated assays.[10] A new strain of entomopathogenic bacterium, *Brevibacillus laterosporus*, was recently isolated from soil in Sardinia (Italy). It showed high toxicity toward house fly adults and larvae[11] and has been screened using LC_{50} estimates on second instar larvae on house fly for management in dairy farms.[12]

Bioassays with fungi are often screened using mean lethal time (LT_{50}) in addition to LC_{50} and LC_{90} because the time necessary to kill a target organism is also relevant for biological control. There are many ways to examine efficacy of entomophathogenic fungi. One method is to first develop concentration response curve using several gradient concentrations of the representative strain to estimate the LC_{50}.[9] Replicated treatments of this dose is then monitored over time.[13,14]

8.5 CONCLUSIONS

Paula Maven offers Hans Gruber several tidbits of advice for his future bioassays with microbials. She reminds him that because of natural variation, responses of a group of insects tested will vary through time, even if all other bioassay techniques and conditions are standardized.[15,16] She advises Dr. Gruber to continue the side-by-side bioassays of the reference standard versus each test sample. Collection of data from side-by-side bioassays permits a continually updated database from which natural variation can be computed.

The potency of each sample can be calculated directly from the relative toxicity value produced by PoloJR.[5] The estimate of potency for the entire lot should be computed from the mean potency estimated from a minimum of four replicated bioassays. The 95% confidence interval should likewise be evaluated for natural variation to ensure that the results are fairly consistent.

Because of natural variation, potencies of both the standard and lots will vary around a mean value. In 95% of the bioassays, the mean will lie within the limits estimated by 1.96 times the SD of the mean. However, 1 in 20 (5%) of the means will be outside the limits simply by chance. Unless a trend is observed over time, a decrease in potency is no cause for alarm.

REFERENCES

1. Hickle, L. A. and Fitch, W. L., eds., *Analytical Chemistry of Bacillus thuringiensis,* ACS Symposium Series 432, American Chemical Society, Washington, DC, 1990.
2. Beegle, C. C., Bioassay methods for quantification of *Bacillus thuringiensis* δ-endotoxin, in Hickle, L. A. and Fitch, W. L., eds., *Analytical Chemistry of Bacillus thuringiensis,* ACS Symposium Series 432, American Chemical Society, Washington, DC, 1990, pp. 14–21.
3. Finney, D. J., *Probit Analysis,* Cambridge University Press, Cambridge, England, 1971.
4. Savin, N. E., Robertson, J. L., and Russell, R. M., A critical evaluation of bioassay in insecticide research: Likelihood ratio tests of dose-mortality regression, *Bull. Entomol. Soc. Am.* 23, 257, 1977.
5. LeOra Software LLC, PO Box 582, Parma, MO 63870. See http://www.LeOra-Software .com.
6. Faust, R. M. and Bulla, L. A. Jr., Bacteria and their toxins as insecticides, In Kurstak, E., ed., *Microbial and viral pesticides,* Marcel Dekker, New York, pp. 75–208, 1982.
7. Russell, T. L., Brown, M. D., Purdie, D. M., Ryan, P. A., and Kay, B. H., Efficacy of VectoBac (*Bacillus thuringiensis* variety *israelensis*) formulations for mosquito control in Australia, *J. Econ. Entomol.* 96, 1789–1791, 2003.
8. Gingrich, R. E. and Eschle, J.L., Susceptibility of immature horn flies to toxins of *Bacillus thuringiensis. J. Econ. Entomol.* 64, 1183–1188, 1971.
9. Lysyk, T. J., Kalischuk-Tymensen, L. D., and Selinger, L. B., Mortality of adult *Stomoxys calcitrans* fed isolates of *Bacillus thuringiensis, J. Econ. Entomol.* 105, 1863–1870, 2012.

10. Maciejewska, J., Chamberlain, W. F., and Temeyer, K. B., Toxic and morphological effects of *Bacillus thuringiensis* preparations on larval stages of the Oriental rat flea (*Siphonaptera: Pulicidae*), *J. Econ. Entomol.* 81, 1656–1661, 1988.

11. Ruiu, L., Floris, I., Satta, A., and Ellar, D. J., Toxicity of a *Brevibacillus laterosporus* strain lacking parasporal crystals against *Musca domestica* and *Aedes aegypti*, *Biol. Control.* 43, 136–143, 2007.

12. Ruiu, L., Satta, A., and Floris, I., Comparative applications of azadirachtin- and *Brevibacillus laterosporus*-based formulations for house fly management experiments in dairy farms, *J. Med. Entomol.* 48, 345–350, 2011.

13. Raga, A. and Sato, M. E., Time-mortality for fruit flies (Diptera: Tephritidae) exposed to insecticides in laboratory, *Arg. Inst. Biol., São Paulo* 73, 73–77, 2006.

14. Lohmeyer, K. H. and Miller, J. A., Pathogenicity of three formulations of entomopathogenic fungi for control of adult *Haematobia irritans* (Diptera: Muscidae), *J. Econ. Entomol.* 96, 1943–1947, 2006.

15. Robertson, J. L., Preisler, H. K., Ng, S. S., Hickle, L. A., and Gelernter, W. D., Natural variation: A complicating factor in bioassays with chemical and microbial pesticides, *J. Econ. Entomol.* 88, 1, 1995.

16. Finney, D. J., *Statistical Methods in Biological Assay*, Griffin, London, 1964.

CHAPTER 9

Pesticide Resistance

Dr. Maven, this book's protagonist, decides to expand her research (much to the exasperation of Jessica) to include small plot trials. Jessica explains that she nor the lab technician simply cannot add fieldwork to their list of duties. After a brief threat of mutiny, Paula hires a farmhand, Todd Hatch, to undertake the responsibilities of tilling, planting, insect scouting, pest management (i.e., insects and weeds), and harvesting. Todd is a recent high school graduate and is attending a local community college, where he is majoring in agronomy. Clearly, she has found the best candidate (and he was cheap). Shortly after planting, scouting reveals Godfather larvae have arrived at their research station. Dr. Maven gives Todd explicit instructions on what should be sprayed (a pyrethroid), at what rate (2 oz per acre), and how often (every 3 weeks for the next 3 months). Not understanding how to convert 2 oz per acre to a rate for a small plot (i.e., 13 ft × 40 ft), Todd sprays 2 oz per plot every 3 days for 3 months. He runs out of the chemical after a month and simply buys two replacement jugs from the local chemical store for $150 each.

During this time, Todd has not seen any larvae in the treated plots and assumes whatever he has been doing is working well. Unbeknown to Todd, a few surviving larvae have crawled off to nearby brush, pupated, and eclosed to adults. These adults fly off, mate, and mature at an astonishing rate (11 days). Two months have passed and local growers have started to call Dr. Maven to complain about spray failures. They want to know if she is breeding some superbugs at the research center. A few growers threaten to sue her department if they find out she is responsible for their superfluous numbers of larvae. Not only that, but Dr. Maven's small plots are full of tiny larvae, her plants look hideous, and most importantly, why does she have a bill for $450 for a cheap pyrethroid?

Paula first asks Jessica to collect as many Godfather larvae as she and her student volunteers can get, from as many sites as possible in Schaefferville and from as many farms outside town as they can visit. Next, she visits Bill Emerine, a consultant, who recounts the details of his scouting horror from a local grower and the wrath he received from his wife for bringing home the larvae and infesting her tomato plants.

During the next few weeks, Jessica and Todd (he is currently removed from field duty) collect Godfather larvae from numerous host species. They find larvae not only on tomatoes (and on parsley, sage, rosemary, and thyme, as well as cilantro) but also on such diverse plant species as plum, pomegranate, pigweed, thistle, and pine. As the insects are collected and the caterpillars reach the proper stage of development, Jessica transfers them to foliage of the same species of plant from which they were collected. After 24 hours, she applies the pyrethroid at a rate equivalent to ounces per acre.[1,2]

As part of her quality control bioassays, Jessica had already estimated baseline susceptibility data from laboratory-reared insects founded with parents collected from Dr. Maven's tomato plants and from several other locations in Schaefferville. When all of the results are in, Dr. Maven discovers that the pyrethroid is not as toxic to Godfather larvae from her yard as it is to caterpillars from other locations, nor is it as toxic as it was to the original groups that Jessica tested when the laboratory colony was established. Paula also notes that larvae on different host plant species collected from the same location seem to differ in response. Responses to the pyrethroid seem to have shifted so that it is generally less toxic and in some collections substantially so.

Dr. Maven has entered a new area of research—pesticide resistance. But before she does anything further, she needs a good working definition of *resistance*.

9.1 RESISTANCE DEFINED

The traditional definition of *resistance*, paraphrased by French-Constant and Roush[3] as "the development of a strain capable of surviving a dose lethal to a majority of individuals in a normal population," has evolved with the completion of increasing research over the past 35 years. Among other problems, a "normal" population response does not exist (see Chapter 6). Ball[4] was probably the first to advocate definition of resistance based on failure of a pesticide to control populations in the field, despite the fact that the same agent adequately controlled the arthropod in the past. This general definition can easily be extended to other chemical or biological materials used to manage a disease, behavior, or another physiological process. Sawicki[5] has proposed an improved definition of *resistance* as "a genetic change in response to selection by toxicants that may impair control in the field." In general, *resistance* is a significant, genetically based shift in the molecular, biochemical, or behavioral bases in populations of an arthropod species, which is measured at the molecular level. *Tolerance*, on the other hand, is a significant difference in quantal responses with which there are varying degrees. Resistance represents one extreme of response, compared with *susceptibility*, the other extreme. In this chapter, the word *tolerance* will be used when discussing methods, results, and interpretation of bioassays.

9.2 NATURAL VARIATION VERSUS TOLERANCE

Tolerance can be distinguished from a major confounding factor—natural variation—by the 95% limits for natural variation of susceptible populations.

Approximate limits of natural variation for susceptible populations from the same host and subjected to identical types of bioassays are mean ± 1.96 (SD), where mean and SD are the mean and standard deviations from the n samples. If the arthropod species is polyphagous, all comparisons should be made from population samples collected from the same host plant species because factors from the plants might also affect the manifestation of resistance (see Section 9.5).

For example, Rust et al.[6] has calculated a mean ± SD LD_{50} of 0.52 ± 0.17 ppm of imidacloprid fed to four susceptible cat flea (*Ctenocephalides felis* [Bouch]) strains tested in three different laboratories. Paula Maven calculates the 95% limits of natural variation for these susceptible cat flea strains to be 0.19–0.85 ppm. Resistance to imidacloprid in cat fleas has not been documented, but these limits are useful for documenting the status of present and future field-collected strains. For example, the LD_{50}s in 7 of 18 bioassays of field-collected cat flea isolates were >0.85; 11 strains can be considered to be susceptible and the 7 other strains are tolerant to imidacloprid. Baseline studies such as this and others are important because they provide documentation both before[2,7–9] and after resistance develops.

9.3 USE OF BIOASSAYS TO SEPARATE POPULATIONS AND STRAINS

Quantal response bioassays are useful to identify and monitor shifts in population tolerance. They provide information that can be used for statistical comparisons of entire regression lines and individual dose levels of interest. Although full-scale bioassays are crucial to toxicological research programs concerning resistance, they are impractical for resistance monitoring programs because of the time and resources involved in their completion. Instead, resistance monitoring programs may use only one dose (i.e., diagnostic or discriminating dose) for screening field populations.

9.3.1 Population Bioassays

Bioassays to compare susceptible populations both among themselves and versus resistant populations should be performed as described in Chapter 3. Likelihood ratio tests (see Section 3.3.1.1) can be used to test whether regression lines are parallel or equal, and the ratio test (see Section 3.3.1.2) can be used to test whether a given LD_x is significantly different from the LD_x of the most susceptible population. The alternative procedure—examination of 95% confidence limit (CL) to detect overlap—must not be used because it lacks statistical power.[10] In general, the lethal concentration (LC) ratio test is more powerful for distinguishing differences between populations than using the CLs around each LC and concluding that differences are significant if there is no overlap.[10] Regression lines of each population can also be plotted to visually assess appropriate groupings of response.[6,7]

Proper dose placement and adequate sample sizes must always be used in all bioassays done to estimate relative tolerance (or resistance) of populations or strains. Sample sizes will dictate the lethal dose (LD) that can reliably be used for comparisons and for fairly reliable estimation of a single dose or set of doses that will discriminate between susceptible and resistant groups. If sample sizes from field-collected populations range from 250 to 500, comparisons at LD_{50} are justified. If ≥750 arthropods are tested in each bioassay, then both LD_{50} and LD_{90} can be used as long as doses are placed as described in Chapter 5. Comparisons at LD_{95} require sample sizes ≥1000; comparisons at LD_{99} require bioassays with 3000–3600 (see Chapter 5); both LDs also require special placement of doses. Unless these very large numbers of subjects are available for bioassays and doses are placed extremely carefully, estimates of these extreme levels are so imprecise that they are meaningless because of their extremely wide confidence intervals.

9.3.2 Response Ratios

Many investigators use response ratios to estimate the magnitude of tolerance or resistance. In much of the literature about resistance, these ratios have been reported as simply as LD (population A)$_x$ ÷ LD (population S)$_x$, where x is the response level, A is the population being compared, and population S is the population with the greatest susceptibility. However, each ratio has an associated error term (see Section 3.2.1.2), and the error must be reported in the form of 95% CLs.[11] Response ratios alone cannot identify resistant populations, but they are useful in the documentation of spatial and temporal differences among populations.

9.3.3 Use of a Discriminating Dose

Use of a single dose for resistance monitoring is more efficient than estimating a complete dose–response curve,[12,13] but the dose or doses must be chosen carefully. Roush and Miller[14] have discussed various aspects of the statistical design of bioassays that would provide reliable estimates. Their premise was that the dose selection and sample size guidelines presented for LD_{90} by Robertson et al.[15] were applicable to more extreme levels such as LD_{95} and LD_{99}. Paula Maven has learned that, although the guidelines for dose selection at the more extreme doses are generally applicable, the guidelines for sample sizes are not (see Section 5.2). Sample sizes for estimation of reliable LD_{95}s are about three times larger than sample sizes required for precise estimation of LD_{90}. For estimation of LD_{99}, at least 30 times more test subjects are required. Collection, handling, and testing of 3000–3600 test subjects in a bioassay done to estimate an extremely high LD is clearly too labor intensive to be done on a routine basis in resistance monitoring (details on discriminating dose selection can be found in Chapter 5).

9.4 STATISTICAL MODELS OF MODES
OF RESISTANCE INHERITANCE

Studies of resistance inheritance usually begin with dose–response bioassays to study aspects of simple Mendelian genetics; these investigations serve as the basis of more sophisticated biochemical and molecular genetic analyses.[16,17] Statistical methods used for genetic analyses must be based on the experimental design of the study *and* the sources of variation in the experiments. Because most bioassays include uncontrollable sources of variation beyond those assumed by the binomial model, these sources of variation must be identified and included in the statistical models. As a guide, Paula will use data analyses from studies of cyhexatin and propargite resistance in Pacific spider mite. With some modifications, similar methods can be adapted to Godfather larvae and other arthropod species.

9.4.1 Standard Method of Analysis with Bioassay Data

Luckily, Paula remembers possible causes of lack-of-fit, use of the χ^2 statistic, and the definition of a residual (see Section 3.2.1.1). Hypotheses about mode of inheritance are tested by using a χ^2 test to compare expected dose–mortality curves calculated from parent (resistant and susceptible) strains and F_1 progeny, versus observed dose–mortality curves of the backcross progeny.[18–20] Significantly large values of the χ^2 statistic may indicate that the putative mode of inheritance is wrong or that the dose placement may require some modification. However, other interpretations are possible, including the presence of extra sources of variability in the bioassay that were not included in the model.[21]

9.4.1.1 Degree of Dominance

Data from dose–response experiments done with susceptible (S), resistant (R), or heterozygous reciprocal F_1 (RS or SR) colonies of arthropods can be analyzed by fitting probit or logit regressions with a standard probit analysis program.[22] The degree of dominance (*D*) of the resistance trait is calculated by the equation

$$D = \frac{2\hat{\theta}_3 - \hat{\theta}_2 - \hat{\theta}_1}{\hat{\theta}_2 - \hat{\theta}_1}, \tag{9.1}$$

where $\hat{\theta}_1 = \log_{10}(LC_{50})$ of the susceptible (S) colony, $\hat{\theta}_2 = \log_{10}(LC_{50})$ of the resistant (R) colony, and $\hat{\theta}_3 = \log_{10}(LC_{50})$ of the heterozygous RS and SR colonies.[23] The standard error of *D* can be calculated based on the asymptotic formula for variances of functions of random variables[24] by taking the square root of

$$\text{var}(D) = \frac{4}{\left(\hat{\theta}_2 - \hat{\theta}_1\right)^2} \cdot \left[\text{var}\left(\hat{\theta}_3\right) + \frac{\left(\hat{\theta}_3 - \hat{\theta}_1\right)^2}{\left(\hat{\theta}_2 - \hat{\theta}_1\right)^2} \text{var}\left(\hat{\theta}_2\right) + \frac{\left(\hat{\theta}_3 - \hat{\theta}_2\right)^2}{\left(\hat{\theta}_2 - \hat{\theta}_1\right)^2} \text{var}\left(\hat{\theta}_1\right) \right]. \tag{9.2}$$

This standard error is used to determine whether *D* is significantly different from ±1 (i.e., whether the resistance trait is completely dominant or completely recessive).

For a completely recessive trait, $D = -1$; for a completely dominant trait, $D = 1$. For example, if $D = -0.87$ and $SE = \sqrt{var(D)} = 0.09$, then the resistance trait may be completely recessive because D is not significantly different from -1. The approximate 95% confidence interval for D ($D \pm 2\,SE = [-1.15, -0.69]$) includes the value -1.

9.4.1.2 Hypothesis Testing

Expected mortalities for the various crosses (SR × R, RS × S, or RS × RS) are then calculated assuming a particular model for mode of inheritance. For example, assuming one major gene and Mendelian mode of inheritance, response probabilities at each concentration for the backcross RS × SS colony (produced by crossing RS females with SS males) should be the average of the mortality probabilities for the RS and SS individuals. The standard χ^2 test is based on the statistic

$$X_i^2 = \frac{(O_i - E_i)^2}{E_i} + \frac{\left(\bar{O}_i - \bar{E}_i\right)^2}{\bar{E}_i}, \tag{9.3}$$

where O_i and E_i are the numbers of observed and expected responses at a given dose d_i and $\bar{O}_i$ and $\bar{E}_i$ are numbers of observed and expected subjects that do not respond. If n_i = total number of subjects from the backcross colony treated with the ith level of toxicant, r_i = observed number of deaths out of n_i subjects, and $\hat{\pi}_i$ = estimated response probability under the proposed genetic model,

$$X_i^2 = \frac{\left(r_i - n_i \hat{\pi}_i\right)^2}{n_i \hat{\pi}_i \left(1 - \hat{\pi}_i\right)}. \tag{9.4}$$

The genetic hypothesis is then tested by comparing the test statistic X_i^2 (for each $i = 1,\dots, N$) with values from a χ^2 table with 1 df or by comparing the sum

$$X^2 = \sum_1^N X_i^2$$

with values from a χ^2 table with N df.

9.4.2 Inferences Using the Standard Method

When the test statistic is not significantly large, Dr. Maven might be tempted to conclude that the hypothesis about mode of inheritance is correct. However, statistical tests can only be used to reject hypotheses, not to accept them. The only conclusion that she can make about a small χ^2 value is that her hypothesis cannot be rejected and that her model describes the data adequately. Other models may also describe the data adequately; such alternative models cannot be rejected unless they are tested individually. If the value of X_i^2 in Equation 9.4 is large and significant, Paula might conclude that her assumptions about the mode of genetic inheritance are incorrect. However, other factors listed next may be the source(s) of a large χ^2 value.

9.4.2.1 Mode of Inheritance

The formula for estimating π_i in Equation 9.4 may be inadequate for at least one dose level. For example, if $\hat{\pi}_i$ were calculated assuming a one-major-gene model, then large values of X_i^2 could mean that the single gene model is inadequate and that some other model (e.g., multiple genes or modifier genes) is necessary to describe the mode of inheritance.

9.4.2.2 Types of Variation

Binomial variation is always present in dose–response experiments. It is that variation in data caused by natural variation in a population. Other sources of variation (e.g., slight differences in temperature and humidity over time, or slight differences in dose between replicates) might also be present. This type of fluctuation has been called *overdispersion*.[25] Experiments in which precise amounts of a material are applied to each test subject, such as topical application to single insects, exhibit less dispersion between replicates than do other types of experiments such as those in which the material is sprayed or incorporated into a diet fed to test subjects.[26]

The χ^2 statistic in Equation 9.3 also assumes that random fluctuations of responses around expected values follow the binomial distribution and that sample sizes are large. The denominator in Equation 9.3 is the variance of binomial variables and the equation includes no other source of variation. Therefore, any source of variation beyond that of the binomial variation will cause the χ^2 statistic to be large even when the hypothesis about the mode of inheritance is not rejected. In most cases, an increase in individuals tested at each dose will reduce the χ^2 statistic.

9.4.2.3 Both Mode of Inheritance and Binomial Distribution

Assumptions about the mode of inheritance and the binomial distribution have the same effect on the χ^2 statistic in Equation 9.3 (i.e., they can cause large values of X_i^2). The first indication of possible overdispersion in bioassay data is a large value of the χ^2 goodness-of-fit statistic for the susceptible (S) or the resistant (R) colonies. The only assumptions involved are that the data fit the probit model and the distribution of responses is binomial. If the probit curve is adequate but the χ^2 goodness-of-fit statistic is still large, then the bioassay data probably show overdispersion.

A second (and probably a more satisfactory method) to detect overdispersion in bioassays with pure R or S colonies or the backcross colonies is to calculate a χ^2 statistic without any assumptions about the "shape" or model of the response probabilities. One such statistic is given by

$$X^2 = \sum_{i=1}^{I} \sum_{j=1}^{J_i} \frac{\left(r_{ij} - n_{ij}\hat{p}_i\right)^2}{n_{ij}\hat{p}_i\left(1 - \hat{p}_i\right)}, \tag{9.5}$$

where the subjects are divided into I dose groups (indexed with i from 1 to I). In the ith dose group, there are J_i subject groups (indexed with j from 1 to J_i), all of which receive the same dose, d_i. In a dose group i, subject group j, there are r_{ij} responses out of n_{ij} subjects (all getting dose d_i); $\hat{p}_i$ is the average fraction of responses at dose d_i—i.e., $\hat{p}_i = (1/J_i)\sum_{j=1}^{J_i}(r_{ij}/n_{ij})$. The only assumption involved is that the responses are independent binomial variables. Therefore, a large χ^2 value will indicate that the data do not consist of independent binomial variables and that the data are overdispersed. Other procedures for testing the assumptions of the binomial distribution are also available.[27]

If Dr. Maven makes an incorrect assumption about the mode of inheritance (e.g., that a one-major-gene model is correct when the mode of inheritance is better described by a multiple gene model), she will see large values of the χ^2 statistic in Equation 9.4, but not extra variation beyond the binomial variation. Consequently, a test statistic such as the one in Equation 9.5 can be used to detect and characterize extra sources of variation that could then be incorporated in the model. One way that she can include overdispersion in a χ^2 test is to calculate a modified statistic

$$X_i^2 = \frac{\left(\bar{r}_i - n_i\hat{\pi}_i\right)^2}{\operatorname{var}\left(\bar{r}_i\right)}, \tag{9.6}$$

where

$$\bar{r}_i = \sum_{j=1}^{J_i} w_{ij} r_{ij}, \tag{9.7}$$

$w_{ij} = n_i/(J_i n_{ij})$, $n_i = \sum_{j=1}^{J_i} n_{ij}$, and where the form for $\operatorname{variable}\left(\bar{r}_i\right)$ depends on the characteristics of the random sources of variation in the particular experiment.

For example, if the only source of variation in the experiment is the binomial fluctuation, then $\operatorname{variable}\left(\bar{r}_i\right)$ can be estimated by $n_i\hat{\pi}_i\left(1 - \hat{\pi}_i\right)$ and the modified statistic in Equation 9.6 is the same as the standard statistic in Equation 9.5 when the sample sizes, n_{ij}, are the same at each replicate. A second form for $\operatorname{variable}\left(\bar{r}_i\right)$ is given in Section 9.4.3.

9.4.2.4 Other Causes for Bad Fit

9.4.2.4.1 Variation between Replicates

Other problems that will result in large values of the χ^2 statistic are outliers and variation between experiments (see Section 3.2.1.1). Except in cases where precise amounts are applied, the actual dose received by each test subject varies even when the amount administered is the same. This source of variation affects the slope and intercept of the fitted probit (or logit) lines. The larger the imprecision, the smaller are the absolute values of the slope and intercept.[28] This source of variation is incorporated in the model automatically because the slope and intercept are estimated from the data; it is not the same as the variation between replicates that is introduced because of differences in dose. Residuals should be examined after each bioassay to determine if any outliers are present. If one or more outliers are found, it is not acceptable to discard only those data points. If the data are truly horrendous (i.e., χ^2 is very high), data from the entire data set for that single date should be excluded with explicit notation recorded as to why the bioassay data was not used.

9.4.2.4.2 Genetic Heterogeneity

In almost all bioassays, the R or S colonies are not totally genetically pure. This problem introduces an additional source of variation that affects the slope of the probit (or logit) line. The larger the heterogeneity is in a colony, the lower the slope of the probit line.[29] In some cases, heterogeneity in a colony will cause a plateau in the line to appear. However, heterogeneity affects the shape of the response curve but not the variation around that curve.

9.4.3 Examples

Hoy et al.[20] and Hoy and Conley[21] have described experiments done to determine modes of inheritance of cyhexatin/fenbutatin oxide and propargite resistances in Pacific spider mite, *Tetranychus pacificus* McGregor. These examples demonstrate the analyses necessary to derive a test statistic or a statistical testing procedure that is appropriate to this set of experiments, and they provide a general guide to statistical analyses for other resistance studies.[30]

9.4.3.1 Dose–Response Bioassays

Both studies involved selection with the acaricide to obtain homogeneous resistant colonies; comparison of dose–mortality regressions of resistant and susceptible colonies, F_1, and backcross progeny; and analysis of observed and expected dose–mortality relationships with standard χ^2 tests. Because this species is arrhenotokous, F_2 females are genetically the same as backcross females. Haploid F_2 males would respond in a 1:1 (R:S) ratio of phenotypes if resistance was conferred by a single major gene and they inherited resistance genes from their mothers. In both studies, two sets of bioassays were done. In the first set, simultaneous bioassays were done for the S, R, SR, and RS colonies. In the second set, simultaneous bioassays were done for the S, R, and backcross SR × S and RS × R colonies.

9.4.3.2 Dose–Mortality Lines

Except for the backcross SR × S and RS × R colonies, dose–mortality data from all bioassays were analyzed with the GLIM[31] statistical package, assuming the probit model and binomial errors. In each experiment, the χ^2 goodness of fit statistic was too large (see Table 9.1), suggesting that the model was not appropriate. A probit line seemed to fit each set of data adequately and the scatter of residuals was apparently random. However, the number of points with an absolute value >2 (see Section 3.2.1.1) suggested overdispersion in the bioassay data of S, R, SR, and RS colonies. Because bioassays of backcross SR × S and RS × R mites were done simultaneously with those of the S and R colonies, overdispersion was also likely to be present. The χ^2 statistic (Equation 9.5) for each of the backcross colonies also

Table 9.1   Goodness of Fit of the Probit (Binomial) Model to Concentration–Response Data for Pacific Spider Mites[a]

Test	Female Colony	Experiment	Number of Mites Treated	Test Statistic for Goodness-of-Fit of Model	df	Statistic for Equality of Probit Lines
Cyhexatin[b]						
	SS	1	600	187.1**	118	
		2	1125	380.8**	157	20**
	RR	1	800	214.7**	158	
		2	1140	276.8**	172	9*
	SR	1	640	201.8**	126	
	RS	1	610	154.6**	120	9*
Propargite[c]						
	SS	1	1431	181.9**	69	
		2	1786	188.9**	98	24**
	RR	1	1559	230.8**	86	
		2	1623	197.7**	90	12**
	SR	1	2042	360.9**	114	
	RS	1	2158	264.9**	122	1.2

Source: Reprinted from Preisler HK, Hoy MA, Robertson JL, *J. Econ. Entomol.,* 83, 1649, 1990.

[a] Likelihood ratio test statistic (df = 2) for overdispersed binomial data.[24]
[b] Data from Ref. 20.
[c] Data from Ref. 21.
* Significant ($P = 0.05$) but not highly significant.
** Highly significant ($P = 0.01$).

**Table 9.2   Goodness-of-Fit Tests for the Binomial Error Model $\left(X_1^2\right)$[a] and the Single
Major Gene (Mendelian) Inheritance Model $\left(X_2^2\right)$[b]**

Females from Backcross	Number of Mites Treated	$\left(X_1^2\right)$ for Binomial Error	df	$\left(X_2^2\right)$ for Single Gene Model	df
Cyhexatin[c]					
SR × S	1300	285*	189	14.0	9
RS × R	1335	315*	186	11.4	11
Propargite[d]					
SR × S	1852	237*	94	42.1*	12
RS × R	2309	199*	166	32.1*	13

Source: Reprinted from Preisler HK, Hoy MA, Robertson JL, *J. Econ. Entomol.*, 83, 1649, 1990.
[a] X_1^2 was calculated using Equation 9.4.
[b] X_2^2 was calculated by summing the values of $X_i X_i^2$ in Equation 9.5 over all i.
[c] Data from Ref. 20.
[d] Data from Ref. 21.
* Highly significant ($P = 0.01$).

indicated overdispersion (see Table 9.2). Because both types of statistical tests were done for each experiment separately, the extra variation observed could not be attributed to variation between experiments. Instead, extra variation between replicates at the same dose was likely. Responses of spider mites from the same replicate thus seemed to be more alike than those from different replicates.

9.4.3.3 Estimation of Overdispersion

Methods for estimation in the presence of overdispersion at the dose level include the beta-binomial method,[32,33] the method of moments,[34,35] and maximum likelihood method.[26] Maximum likelihood procedures were used to compute the mortality probability estimates $\hat{P}^{(R)}, \hat{P}^{(S)}, \hat{P}^{(RS)}$, and $\hat{P}^{(SR)}$, and corresponding LC_{50}s and standard errors. This procedure incorporates extra variation into the model by adding a random effect variate to the fixed effect regressors at the probit scale. The same procedure can also be used to test hypotheses about equality of probit lines. When the equality of the probit lines for the F_1 colonies (SR and RS colonies) was tested in the cyhexatin and propargite studies, differences in the probit lines were not highly significant (see Table 9.1). Therefore, data from the two F_1 colonies were pooled and an estimate for $P^{(F_1)}$ was obtained from the pooled data.

When tests of equality of probit lines were done for the S and R colonies from experiments with cyhexatin and propargite, probit lines from the various experiments were highly significant in almost all cases (see Table 9.1). Preisler et al.[25] then attempted to incorporate the variability between experiments into the model by fitting a model with a probit line

$$\alpha_k + \beta_k \log(d) + \delta_l, \tag{9.8}$$

where k is the indicator for the various colonies and l is the indicator for the various experiments. In this model, the experiment effect δ_l is incorporated as an additive effect on the probit scale. This model seemed to fit the data from the propargite studies but not those from the cyhexatin studies. A further search for an appropriate model describing the variation between experiments in the cyhexatin studies was not done.

9.4.3.4 Mode of Inheritance of Cyhexatin Resistance

Resistance to cyhexatin in Pacific spider mite was incompletely recessive ($D \pm$ SE $= -0.18 \pm 0.05$ calculated with the pooled F_1 data). Expected response probabilities for the backcross females (RS × R and SR × S) were calculated assuming that resistance to cyhexatin is determined by a single major gene with Mendelian mode of inheritance—that is,

$$\pi_i = 0.5\hat{P}_i^{F_i} + 0.5\hat{P}_i^{(R)} \tag{9.9}$$

for the RS × R backcross colony and

$$\hat{\pi}_i = 0.5\hat{P}_i^{F_i} + 0.5\hat{P}_i^{(S)} \tag{9.10}$$

for the SR × S backcross colony. The standard χ^2 test statistics, calculated with Equation 9.4, were too large in each case. This result was not surprising because the χ^2 statistic (assuming binomial error) was too large even when a nonparametric formula (Equation 9.5) was used to estimate π_i. Because of the overdispersion detected in these data, Equation 9.6 was used with a variance function that incorporates the extra variation to calculate a modified test statistic. One simple way to incorporate the overdispersion in the variance function is to use the formula

$$\mathrm{var}\left(\bar{r}_i\right) = n_i\hat{\pi}_i\left(1 - \hat{\pi}_i\right)h, \tag{9.11}$$

where

$$h = \frac{X^2}{\sum_i J_i - I}$$

and where X^2 is calculated using Equation 9.5. The value h is an estimate of the heterogeneity factor.[29] Equation 9.10 is a reasonable estimate for the variance of $\bar{r}_i$ when the number of subjects n_{ij} treated at each replicate are equal. A more

appropriate formula for the case when the sample sizes at each replicate are not equal
is given by

$$\text{var}(\bar{r}_i) = \frac{n_i^2}{J_i^2}\left[m_i\hat{\pi}_i\left(1 - \hat{\pi}_i\right) + (J_i - m_i)\hat{\sigma}_i^2 \right], \qquad (9.12)$$

where

$$w_{ij} = n_i/(J_i n_{ij}), m_i = \sum_j (1/n_{ij})$$

and where $\hat{\pi}_i$ and $\hat{\sigma}_i^2$ are estimated by fitting probit curves to the R, S, RS, and SR
mortality data as described by Preisler.[36]

Values of the modified χ^2 statistics for the cyhexatin experiments calculated with
Equations 9.6, 9.7, and 9.12 are listed in Table 9.2. These values seem to indicate no
significant departure of the expected responses from the single major gene model as
given by Equations 9.9 and 9.10.

9.4.3.5 Mode of Inheritance of Propargite Resistance

Values for degrees of dominance indicated that resistance to propargite was
incompletely recessive ($D \pm SE = -0.30 \pm 0.03$, calculated with the pooled F_1 data).
Modified χ^2 values, assuming a single major gene Mendelian model, were large
(see Table 9.2). When the modified X_i^2 values were examined individually, 4 of the
12 values from the SR × S cross and 2 of the 13 values from the RS × R cross were
significant ($P < 0.01$). In each case, responses of Pacific spider mites to propargite
in the backcross colonies seemed significantly larger than expected. Reasons for
this significant difference between observed and expected are not clear. Possibly,
the assumption that a model with one major gene describes the expected responses
was not correct. The mode of inheritance of propargite resistance might be better
explained by a polygenic model or a model with modifier genes. Also, the model for
incorporating the variation between experiments (Equation 9.10) might not be appropriate for the RS and SR colonies. However, this possibility could not be checked
because the bioassays on these colonies were done only in one set of experiments.

9.5 HOST–INSECT INTERACTION
AND THE EXPRESSION OF RESISTANCE

Factors that affect the phenotypic expression of genetic resistance might affect
not only the results of bioassays done in toxicological experiments but also the results
of resistance detection with discriminating doses. Robertson et al.[37] examined levels of nonspecific esterases and glutathione-S-transferase in resistant and susceptible strains of light brown apple moth, *Epipyas postvittana* (Walker), last instars
that fed on different host plant species. This insect species, which is resistant to

the organophosphorous pesticide azinphosmethyl, is polyphagous and feeds on such diverse hosts as gorse and broom, in addition to apple. Although higher glutathione-S-transferase activities in resistant insects fed blackberry, apple, gorse, broom, or artificial diet may be responsible for decreased toxicity of azinphosmethyl in the resistant strain, an allelochemical in blackberry foliage may inhibit or prevent induction of nonspecific esterase activity in the same strain. In resistance detection in the field, resistant larvae off of blackberry might mistakenly be identified as susceptible when they are not. These results indicate the need for a careful and additional exploration of the factors, including those related to host plants that affect the phenotypic expression of resistance.

9.6 INSECT GROWTH REGULATORS AND RESISTANCE

Resistance can be expressed in any developmental stage of an insect, so it is no surprise that various insect species have started to express lower susceptibility to these chemicals. For example, laboratory colonies of *Culex pipiens pipiens* L., *Tribolium conjusum* Jacquelin duVal, and *Oncopelius faciatus* (Dallas) submitted to resistance pressure from juvenile hormone (JH) mimics expressed polyfactoral inheritance and cross-resistance to five other JH mimics.[38]

Cross-resistance has also been suggested between insect growth regulators (IGRs), carbamates, and organophosphates. The tufted apple bud moth, *Platynota idaeusalis* (Walker), was found tolerant to diflubenzuron, hexaflumuron, and the organophosphate, azinphosmethyl.[39] Populations of codling moth (*Cydia pomonella* (L.) (Lepidoptera: Tortricidae) from a commercial apple orchard were found tolerant to fenoxycarb, teflubenzuron, and phosalone.[40] However, this tolerance appears to be species specific. Overexpression of cytochrome P450 gene *Cyp12a4* was found to be the cause of resistance to lufenuron in the fruit fly, *Drosophila melanogaster* Meigen,[41] while populations of the armyworm *Spodoptera litura* (F.) in Pakistan have shown very little tolerance to lufenuron despite intensive use for control of this pest.[42]

Estimation of baseline susceptibility for Lepidopteran larvae can be especially problematic when working with insect growth regulators. Conventional neurotoxins kill susceptible individuals within 24–48 hours with a single recording of mortality. However, the effects from a growth regulator occur only during a molt. As a larva is successful with each future molt, it is less affected by the poison. Jones et al.[43] shifted their bioassays from larval to eggs when incomplete molting at 4 or 7 days produced lower mortality at higher doses. Larvae evaluated 48 hours after treatment[42] or noting successful emergence as adult moths[44] are two additional methods. When two life stages are able to be compared, results from one stage can significantly demonstrate differences in management practices. In the case of sweet potato whitefly (Homoptera: Aleyrodidae), resistance ratios were much larger (554-fold) for egg-hatch than adult emergence (10-fold).[44]

Another factor that may contribute to difficulty in screening for resistance to IGRs is high innate variability in response.[25] This variability occurs in every bioassay. When data are entered into a software program, it does not necessarily mean that the output is high quality. For example, a researcher tests $n = 30$ individuals at each of seven concentrations. This n for each concentration has produced excellent regression lines for several bioassays that the researchers have conducted in the past. However, when the 95% confidence intervals are examined for the present data set, the width is very large or imprecise. The latter is why when estimating LD_{50} from a field population, a sample size of 250–500 individuals should be tested. If LC_{50} and LC_{90} are requested, then ≥ 750 insects should be tested to screen for resistance (Section 9.3.1). The error term associated with the resistance ratios (see Section 3.2.1.2) must be reported in the form of 95% CLs and not as standard errors.[11]

9.7 GENETICALLY MODIFIED CROPS

Transgenic crops that produce toxins from the bacterium *Bacillus thuringiensis* (*Bt*) grew exponentially in the late 1990s, with widespread and prolonged exposure to *Bt* toxins representing one of the largest selection pressures for insect resistance. Transgenic *Bt* maize that produces lower than optimal high dose and lack of compliance with regard to refuge retention presents even more challenges to integrated resistance management. At least seven resistant laboratory strains of three pests (*Plutella xylostella* [L.],[45] *Pectinophora gossypiella* [Saunders],[46,47] and *Helicoverpa armigera* [Hübner][48]) have completed development on *Bt* crops. In contrast, monitoring field populations of *P. gossypiella* and *H. armigera* in regions with high adoption of Bt crops for 5 and 3 years, respectively, has not yet detected increases in resistance frequency.[46–49]

Early detection and mitigation of resistance through rapid diagnostic bioassays are necessary tools to reduce widespread economic losses. Unfortunately, the increase in alleles conferring resistance to *Bt* has become so common, that high-dose resistance strategy[50] is also failing. When the beetle *Chrysomela tremulae* Fabricius fed on *Bt* poplar clone *Poplus tremula* L. × *Populus tremuloides* Michx with Cry3Aa toxin equal to $6.34 \times LC_{99}$ of the F1 generation, the resistance ratio of the F2 offspring was >6200 and almost identical to the nontransgenic poplar plants.[51] Backcrosses between susceptible, resistance, and F1 hybrids showed that resistance was conferred by a single autosomal gene and was almost completely recessive ($D_{LC} = 0.07$).

Susceptibilities of insects to Cry proteins can be developed and used as a baseline for screening field populations. However, various manufacturers of these proteins can generate strains that vary in their toxicity; hence, LC_{50} values can also vary significantly.[52] It is a wise decision that once a manufacturer of a toxin strain is selected, the particular manufacturer is also used in the future. Otherwise, changes in *Bt* batches may be mistaken as resistance. Likewise for test procedures. If diet incorporated assays are selected rather than residual or surface treatment, future monitoring should replicate the baseline procedures. However, if methods must be adapted, comparison between these two test procedures for European corn borer, *Ostrinia nubilalis* (Hübner), has shown no significant difference with regard to their practicability and efficiency.[52]

REFERENCES

1. Robertson, J. L. and Rappaport, N. L., Direct, indirect, and residual toxicities of insecticide sprays on western spruce budworm, *Choristoneura occidentalis* (Lepidoptera: Tortricidae), *Can. Entomol.* 111, 1219, 1979.

2. Bergh, J. C., Rugg, D., Jansson, R. K., McCoy, C. W., and Robertson, J. L., Monitoring the susceptibility of citrus rust mite (Acari: Eriophyidae) populations to abamectin, *J. Econ. Entomol.* 92, 781, 1999.

3. French-Constant, R. H. and Roush, R. T., Resistance detection and documentation, in Roush, R. T. and Tabashnik, B. E., eds., *Pesticide Resistance in Arthropods*, Chapman & Hall, NY, 1990, pp. 4–38.

4. Ball, H. J., Insecticide resistance: A practical assessment, *Bull. Entomol. Soc. Amer.* 27, 261, 1981.

5. Sawicki, R. M., Definition, detection and documentation of insecticide resistance, in Ford, M. G., Holloman, D. W., Khambay, B. P. S., Sawicki, R. M., eds., *Combating Resistance to Xenobiotics; Biological and Chemical Approaches*, Ellis Harwood, Chichester, England, 1987, pp. 105–117.

6. Rust, M. K., Denholm, I., Dryden, M. W., Payne, P., Blagburn, B. L., Jacobs, D. E., Mencke, N., Schroeder, I., Vaughn, M., Mehlhorn, H., Hinkle, N. C., and Williamson, M., Determining a diagnostic dose for imidacloprid susceptibility testing on field-collected isolates of cat fleas (Siphonaptera: Pulicidae), *J. Med. Entomol.* 42, 631, 2005.

7. Pree, D. J., Archibald, D. E., Ker, K. W., and Cole, K. J., Occurrence of pyrethroid resistance in pear psylla (Homoptera: Psyllidae) populations in southern Ontario, *J. Econ. Entomol.* 83, 2159, 1990.

8. Scott, J. G., Roush, R. T., and Rutz, D. A., Insecticide resistance of house flies from New York dairies (Diptera: Muscidae), *J. Agric. Entomol.* 6, 53, 1989.

9. Staetz, C. A., Susceptibility of *Heliotis virescens* (F.) (Lepidoptera: Noctuidae) to permethrin across the cotton belt: A five year study, *J. Econ. Entomol.* 78, 505, 1985.

10. Wheeler, M. W., Park, R. M., and Bailer, A. J., Comparing median lethal concentration values using confidence interval overlap or ratio tests, *Environ. Toxicol. Chem.* 25, 1441, 2006.

11. See Robertson, J. L., Preisler, H. K., Ng, S. S., Hickle, L. E., and Gelernter, W. D., Natural variation: A complicating factor in bioassays with chemical and microbial pesticides, *J. Econ. Entomol.* 88, 1, 1995.

12. Sawicki, R. M., Denholm, I., Forrester, N. W., and Kershaw, C. D., Present insecticide-resistance management strategies in cotton, in Green, M. B. and de B. Lyon, D. J., eds., *Pest Management in Cotton*, Ellis Harwood, Chichester, England, 1989, pp. 21–43.

13. Forrester, N. W., Cahill, M., Bird, L. J., and Layland, J. K., Management of pyrethroid and endosulfan resistance in *Helicoverpa armigera* (Lepidoptera: Noctuidae) in Australia, *Bull. Entomol. Res. Suppl.* 1, 1990, 160–168.

14. Roush, R. T. and Miller, G. L., Considerations for design of insecticide resistance monitoring programs, *J. Econ. Entomol.* 79, 293, 1986.

15. Robertson, J. L., Smith, K. C., Savin, N. E., and Lavigne, R. J., Effects of dose selection and sample size on precision of lethal dose estimates in dose-mortality regression, *J. Econ. Entomol.* 77, 833, 1984.

16. Hemingway, J. and Ranson, J., Insecticide resistance in insect vectors of human disease, *Ann. Rev. Entomol.* 45, 371, 2000.

17. French-Constant, R., Anthony, N., Aronstein, K., Rocheleau, T., and Stilwell, G., Cyclodiene insecticide resistance: From molecular to population genetics, *Ann. Rev. Entomol.* 45, 449, 2000.

18. Halliday, W. R. and Georghiou, G. P., Inheritance of resistance to permethrin and DDT in the southern house mosquito (Diptera: Culicidae), *J. Econ. Entomol.* 78, 762, 1985.

19. Roush, R. T., Combs, R. L., Randolph, T. C., Macdonald, J., and Hawkins, J. A., Inheritance and effective dominance of pyrethroid resistance in the horn fly (Diptera: Muscidae), *J. Econ. Entomol.* 79, 1178, 1986.

20. Hoy, M. A., Conley, J., and Robinson, W., Cyhexatin and fenbutatin-oxide resistance in Pacific spider mite (Acari: Tetranychidae): Stability and mode of inheritance, *J. Econ. Entomol.* 81, 57, 1988.

21. Hoy, M. A. and Conley, J., Propargite resistance in Pacific spider mite (Acari: Tetranychidae): Stability and mode of inheritance, *J. Econ. Entomol.* 82, 11, 1989.

22. LeOra Software, *PoloJR*, LeOra Software, PO Box 562, Parma, MO 63870 See http://www.LeOra-Software.com.

23. Stone, B. F., A formula for determining degree of dominance in case of monofactorial inheritance of resistance to chemicals, *Bull. W.H.O.* 38, 325, 1968.

24. Lehmann, E. L., *Testing Statistical Hypotheses*, Wiley, New York, 1966.

25. Preisler, H. K., Hoy, M. A., and Robertson, J. L., Statistical analyses of modes of inheritance for pesticide resistance, *J. Econ. Entomol.* 83, 1649, 1990.

26. Preisler, H. K. and Robertson, J. L., Analysis of time-dose-mortality data, *J. Econ. Entomol.* 82, 1534, 1989.

27. Mosteller, F. and Tukey, J. W., The uses and usefulness of binomial probability paper, *J. Am. Stat. Assoc.* 44, 174, 1949.

28. Burr, D., On errors-in-variables in binary regression—Berkson case, *J. Am. Stat. Assoc.* 83, 739, 1988.

29. Finney, D. J., *Probit Analysis*, 3rd ed., Cambridge University Press, Cambridge, England, 1971.

30. Brun, L. O., Suckling, D. M., Roush, R. T., Caudichon, V., Preisler, H., and Robertson, J. L., Genetics of endosulfan resistance in *Hypothenemus hampei* (Coleoptera: Scolytidae): Implications for mode of sex inheritance, *J. Econ. Entomol.* 88, 470, 1995.

31. Payne, C. D., ed., *The GLIM System Release 3.77 Manual*, Numerical Algorithms Group, Oxford, England, 1978.

32. Williams, D. A., The analysis of binary responses from toxicological experiments involving reproduction and teratogenicity, *Biometrics,* 31, 949, 1975.

33. Crowder, M. J., Beta-binomial Anova for proportions, *Appl. Stat.* 27, 34, 1978.

34. Williams, D. A., Extra-binomial variation in logistic linear models, *Appl. Stat.* 31, 144, 1982.

35. Moore, D. F., Modeling the extraneous variance in the presence of extra-binomial variation, *Appl. Stat.* 36, 8, 1987.

36. Preisler, H. K., Assessing insecticide bioassay data with extra-binomial variation, *J. Econ. Entomol.* 81, 759, 1988.

37. Robertson, J. L., Armstrong, K. F., Suckling, D. M., and Preisler, H. K., Effects of host plants on toxicity of azinphosmethyl to susceptible and resistant light brown apple moth (Lepidoptera: Tortricidae), *J. Econ. Entomol.* 83, 2124, 1990.

38. Brown, T. M., DeVries, D. H., and Brown, A. W. A., Induction of resistance to insect growth regulators. *J. Econ. Entomol.* 71:2, 223–229, 1978.

39. Biddinger, D. J., Hull, L. A., and McPheron, B. A., Cross-resistance and synergism in azinphosmethyl resistant and susceptible strains of tufted apple bud moth (Lepidoptera: Tortricidae) to various insect growth regulators and abamectin. *J. Econ. Entomol.* 89:2, 274–287, 1996.

40. Stará, J. and Kocourek, F., Insecticidal resistance and cross-resistance in populations of *Cydia pomonella* (Lepidoptera: Tortricidae) in Central Europe. *J. Econ. Entomol.* 100:5, 1587–1595, 2007.

41. Bogwitz, M. R., Chung, H., Magic, L., Rigby, S., Wong, W., O'Keefe, M., McKenzie, J. A., Batterham, P., and Daborn, P. J., *Cyp12a4* confers lufenuron resistance in a natural population of *Drosophila melanogaster*. *PNAS* 102, 12807–12812, 2005.

42. Ahmad, M. and Mehmood, R. Monitoring of resistance to new chemistry insecticides in Spodoptera litera (Lepidoptera: Noctuidae) in Pakistan. *J. Econ. Entomol.* 108:3, 1279–1288, 2015.

43. Jones, M. M., Robertson, J. L., Weinzierl, R. A., Susceptibility of eggs from two laboratory colonies of Oriental fruit moth (Lepidoptera: Tortricidae) to novaluron. *J. Agri. Urban Entomol.* 26:4, 175–181, 2011.

44. Horowitz, A. R. and Ishaaya, I., Managing resistance to insect growth regulators in the sweetpotato whitefly (Homoptera: Aleyrodidae). *J. Econ. Entomol.* 87:4, 866–871, 1994.

45. Tabashnik, B. E., Finson, N., Johnson, M. W., and Moar, W. J., Resistance to toxins from *Bacillus thuringiensis subsp. Kurstaki* causes minimal cross-resistance to *B. thuringiensis subsp. Aizawai* in diamondback moth (Lepidoptera; Plutellidae). *Appl. Environ. Microbiol.* 59, 1332–1335, 1993.

46. Tabashnik, B. E., Patin, A. L., Dennehy, T. J., Liu, Y. B., Carriere, Y., Sims, M. A., and Antilla, L., Frequency of resistance to *Bacillus thuringiensis* in field populations of pink bollworm. *Proc. Natl. Acad. Sci. USA,* 97, 12980–12984, 2000.

47. Tabashnik, B. E., Roush, R. T., Earle, E. D., and Shelton, A. M., Resistance to Bt toxins. *Science,* 287, 42, 2000.

48. Fan, X., Zhao, J. Z., Fan, Y., and Shi, X., Inhibition of transgenic Bt plants to the growth of cotton bollworm. *Plant Protection,* 26, 3–5, 2000.

49. Wu, K., Guo, Y., Lv, N., Greenplate, J. T., and Deaton, R., Resistance monitoring of *Helicoverpa armigera* to *Bacillus thuringiensis* insecticidal protein in China. *J. Econ. Entomol.* 95, 826–831, 2002.

50. Georghiou, G. P., and Taylor, C. E., Operational influences in the evolution of insecticide resistance. *J. Econ. Entomol.* 70, 653–658, 1977.

51. Augustin, S., Courtin, C., Rejasse, A., Lorme, P., Genissel, A., and Bourguet, D., Genetics of resistance to transgenic *Bacillus thuringiensis* Poplars in *Chrysomela tremulae* (Coleoptera: Chrysomelidae). *J. Econ. Entomol.* 73, 1058–1064, 2004.

52. Saeglitz, C., Bartsch, D., Eber, S., Gathmann, A., Priesnitz, K. U., and Schuphan, I., Monitoring the Cry1Ab susceptibility of European corn borer in Germany. *J Econ Entomol.* 99, 1768–1773, 2006.

CHAPTER **10**

Mixtures

A few weeks after the arthropod resistance scare, Paula Maven takes an afternoon off from work as a research entomologist to tend to her vegetable garden. The tomato plants look horrible. The leaves look as though they have been riddled with potshot. A closer observation finds a Godfather larva happily munching on a leaf. Paula plucks the worm off the plant and tosses it into a bucket of soapy water. She plucks another, and another, and another. Over the fence, she notices that Molly Weasley is also gardening. Miss Weasley had planted her tomatoes at the same time Paula had, but the leaves of her plants are undamaged. Dr. Maven just has to find out why Molly's plants have escaped damage.

"Miss Weasley, why do your tomatoes look so healthy? I don't see any signs of insect damage, and I was just wondering…"

"Oh, it's simple. I've been spraying them with the chemical you told me to use, but for good luck I add just a pinch of another special white powder that my late brother Fred left in the shed." "Why, he even showed my nephews how to mix that white powder into paint." Fred also owned a house painting business along with his sons Fred and George. "Well, anyway, they were very successful painters, and fast, too." Fifteen stories later, Dr. Maven politely excuses herself and flees to the quiet of her kitchen. The thought of living next door to Fred's shed filled with mysterious white powders is disconcerting; what a toxic waste dump!

Paula now considers the merits of adding a pinch of one chemical to another. Perhaps the idea just might work. She has identified several pesticides that seem both environmentally safe and sufficiently effective to save Griffendor's tomatoes from the ravenous appetites of the Godfather larvae. Of all the candidates, she prefers a diamide because they are most specific to the target. But these newer chemistries are very expensive, perhaps prohibitively so for most local growers.

Dr. Maven now wonders what would happen if a trace amount of the diamide were mixed with a less expensive (but also, less effective) type of pesticide. Would toxicity increase significantly? Another alternative might be adding a synergist— that is, a chemical that is not toxic in itself but that blocks one of the steps involved in detoxification of a pesticide. As a result, toxicity of the pesticide would also increase significantly.

Experiments with synergists are simple, as are data analyses. However, practical aspects of the problem of mixtures of one component with another have been unnecessarily complicated by descriptions of elaborate statistical models developed without biochemical information, let alone any means of experimental verification. In her review of the subject, Dr. Maven notices that these elaborate statistical methods have rarely been used in biological investigations, most likely because they rely on assumptions that are impossible to verify and because they are founded on unsubstantiated theories of insect toxicology. Paula chooses to examine the simple model proposed by Bliss.[1]

Bliss developed the first statistical models of joint action of pesticide mixtures based on three theoretical modes of physiological interaction. These types of interaction are independent joint action, similar joint action, and synergistic action. Efforts to extend and elaborate upon Bliss's categories resulted in development of increasingly complex mathematical formulas to describe physiologically more complex situations.[2–12] As a demonstration of what can happen when statistical theory is developed in the absence of biological or biochemical data, the resultant statistical methods are of limited practical value to biologists.

10.1 INDEPENDENT, UNCORRELATED JOINT ACTION OF PESTICIDE MIXTURES

In a reexamination of the mixture problem, Robertson and Smith[13,14] described statistical methods that can be used to test a very simple null hypothesis first defined by Bliss.[1] This hypothesis—independent, uncorrelated joint action—means that the toxicity of each component in the mixture is not affected by toxicity of the other component and that susceptibility to one mixture component is not correlated with susceptibility to the other. A response will occur only if the concentration of at least one of the two chemicals exceeds the tolerance of the insect. Biochemically, this type of response will be observed if the action sites of the two chemicals or components are different (i.e., if each chemical has a completely different mode of action and these modes of action are totally independent).

Assuming that this type of interaction occurs, synergism is the occurrence of significantly greater mortality than that predicted by the model. The opposite effect, significantly reduced mortality, is defined as antagonism. These definitions are implicit in the original definition of synergistic action given by Bliss[1]: "The effectiveness of the mixture cannot be assessed from that of the individual ingredients but depends upon a knowledge of their combined toxicity when used in different proportions. One component synergizes or antagonizes the other."

10.1.1 Statistical Model

In experiments with a mixture of two chemicals, insects can die of three possible causes. The first cause is natural mortality, with a probability p_0 (a constant). The other two causes of mortality are chemical or component 1 and chemical or component 2. For the first component, the probability of mortality or response (p_1) is a function of dose d_1. Usually, the probit or logit of $X_1 = \log(d_1)$. For the second component, the probability of mortality or response (p_2) is a function of d_2. If these three causes of mortality are independent, the probability of insect death (p) resulting from the mixture of pesticides is

$$p = p_0 + (1 - p_0)p_1 + (1 - p_0)(1 - p_1)p_2,$$

where each plus sign means "or" and each product means "and." This equation can be interpreted as follows: The total probability of death equals death from natural causes (p_0) or no death from natural causes ($1 - p_0$). The other options are no death from natural causes, but death from component 1 (P_1) [e.g., $(1 - P_0)P_1$] or no death from natural causes or component 1 (P_1) [e.g., $(1 - p_0)(1 - p_1)$] and death caused by component 2 (P_2) [i.e., $(1 - p_0)(1 - p_1)p_2$].

10.1.2 Test of Hypothesis of Independent Joint Action

A χ^2 statistic can be used to test the hypothesis of independent joint action. This test is done by obtaining an estimate for the probability of mortality (p) for several dose or concentration levels of the two pesticides and then comparing $\hat{p}$ (the estimate of p) with the observed proportion killed at the corresponding dose levels. The three contributions to p are estimated separately. First, p_0 is calculated as the proportional mortality observed in the control group. Next, p_1 and p_2 are estimated from separate bioassays of pesticide 1 and pesticide 2, with data analyzed using a computer program such as POLO,[15] POLO-PC,[16] PoloPlus,[17] R,[18] or PoloSuite.[19] Finally, the χ^2 test statistic is calculated with MIX,[14] PoloMix,[20] or PoloMixture,[21] the latter two computer programs written specifically for this purpose.

10.1.3 PoloMixture[21]

Microsoft Windows and Apple OS-compatible program PoloMixture[21] was designed to identify pairs of chemicals whose interaction results in a response that departs from the hypothesis predicted by the model of independent, uncorrelated joint action. The program is based on the general procedure described in the previous section and the corrected procedure originally described by Robertson and Smith.[13,14]

Parameter estimates from statistical analyses with data for each component of the mixture and dose–response data for the mixture are put in three data files for use by the PoloMixture program. These parameter files are used to compare and contrast the estimated regression lines of the individual components to that of the line generated by the mixture bioassay. Parameter files are automatically generated by the PoloJR[19] program when a dose–response curve is estimated. These parameters include intercept, estimated slope, estimated variance of the intercept, estimated variance of the slope, estimated covariance of the slope and intercept, and the heterogeneity adjustment factor.

Parameter information for two components, chlorantraniliprole and esfenvalerate, respectively, are as follows. Again, this information is strictly for demonstration as PoloJR automatically generates these files for input into PoloMixture.

Parameter Data for component 1

chlorantraniliprole
4.69 4.23 0.14 0.13 0.12 2.42

Parameter Data for component 2

esfenvalerate
1.05.29 0.009 0.047 0.009 1.25

10.1.3.1 Program Input

The mixture file itself should be a text file (see the following). The first line is for identification only. The actual data follow on subsequent lines. Each line has three columns. The first is dose (x), the second is the number (n) of test subjects, and the third is the number responding (r). The number of control subjects and their responses are entered directly into the program (Figure 10.1).

OFM, chlorantraniliprole, esfenfalerate

0.10 88 15
0.20 72 35
0.30 73 56
0.40 71 60
0.60 64 62

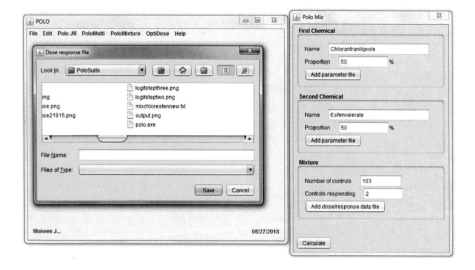

Figure 10.1 The opening PoloMixture window that Jessica has filled in with the information on the two chemicals and the parameter files.

10.1.3.2 Running PoloMixture

Jessica first opens the PoloJR module and opens each individual component data file to estimate its dose–response curve. Following each calculation, she clicks on "generate parameter file" and saves the text file to her "Mixture" directory. Next, she opens the PoloMixture module and fills in the name and ratio or percentage of each component used in the mixture. She selects their parameter files that were generated by PoloJR. In the "Mixture" section, she enters the number of control subjects and the number of individuals that responded. She uploads the mixture file that represents the estimated dose–response curve for the combined two components. The latter file is estimated using PoloJR and should have a minimum of five dose concentrations that produce between 20% and 90% mortality. These are constraints for the PoloMixture program to reliably compare slopes between the lines. She clicks the "Compute" button and the analysis appears. She saves it to a "Results" file (see Figure 10.2).

10.1.3.3 Program Output

PoloMixture calculations of expected mortality and χ^2 values for each dose as described earlier in this chapter (Section 10.1.1) are listed in the last two columns of lines for the mixture doses. Data for the actual bioassay done with the mixture of components 1 and 2 and the total χ^2 for the appropriate degrees of freedom (i.e., number of doses including the control, minus 1) are listed next. To determine whether significant departures from the null hypothesis occur, the critical tabular value of χ^2 at the probability level of choice (e.g., $P = 0.05$) is compared to the calculated χ^2. If the *PoloMixture* value is less than the tabular value, the null hypothesis cannot be rejected. Conversely, if the calculated χ^2 value is greater than the tabular value, then the hypothesis of independent joint action is rejected.

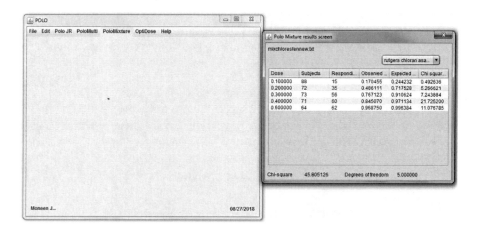

Figure 10.2 PoloMixture results for this set of data.

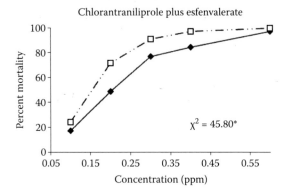

Figure 10.3 Graph of the PoloMixture results for this set of data. Dotted line represents expected mortality and solid line represents observed mortality.

When a hypothesis is rejected, a plot of the data should be examined to determine the direction of significant departures from values predicted by the hypothesis. Ideally, all the values will lie either to the left (significant antagonism) or to the right (significant synergism) of expected values. However, points on either side of the expected line occur when the expected line is not equal or parallel to the observed line; in this situation, χ^2 values for the parts of the line above and below the 50% response level can be examined.[13] If significance does occur, compare the observed and expected mortality columns to determine whether it was synergism or antagonism has occurred. If the majority of the observed is less than the expected, antagonism is suspected, and vice versa. In Figure 10.3, antagonism has occurred when the two chemicals were combined in equal ratios.

10.2 SIMILAR (ADDITIVE) JOINT ACTION

A second hypothesis that can be tested about the joint action of chemicals in a mixture is simple similar action. When each of the two components is toxic and their dose–response lines are parallel, their joint action is additive if dose d_1 of component 1 has the same effect as dose ρd_2 of component 2, where ρ is the relative potency of the pesticides. Such a relationship could occur if the two components are closely related structurally but simply have different potencies.

An additive relationship thus suggests that treating an organism with a mixture d_1 units of pesticide 1 and d_2 units of component 2 is the same as treating it with $d_1 + \rho d_2$ units of component 1 alone. A general parametric equation for similar action is here

$$y = \alpha + \beta \log(d_1 + \rho d_2),$$

where y is the probit or logit line. Diagnostic tests for similar action compared with alternatives corresponding to antagonistic or synergistic joint action are described by Giltinan et al.[22] However, no computer programs are presently available for doing these tests.

10.3 OTHER THEORETICAL HYPOTHESES
OF JOINT ACTION OF PESTICIDES

Finney[23] reconsidered Bliss's three types of interaction and added some minor modifications. Hewlett and Plackett[2-4] have contributed to several new theories of interaction that are so subtle in their differences that they probably cannot be demonstrated biologically. Sakai[24] subsequently expanded the types of chemical interactions described by Hewlett and Plackett to include the following possible types of joint action: true, similar (similar and complex), dissimilar (dependent and independent), compound (similar and dissimilar), actuated, offset, converted, synergistic, antagonistic, pseudo-, physiological (similar and dissimilar), physicochemical, pseudosynergistic, and pseudo-antagonistic. Dr. Maven agrees with Busvine's[25] comment made in reference to Sakai's[24] modifications: "His latest contribution displays a subtlety that may deter many insect toxicologists (including the writer)."

10.4 SYNERGISTS

As mentioned previously, synergists are chemicals that are not toxic to an arthropod. When mixed with a pesticide, however, they block a step of the detoxification of the pesticide so that the pesticide becomes significantly more toxic. For example, Corbett et al.[26] list four chemicals that continue to be of commercial interest. They are N-(2-ethylhexyl)-8,9, 10-trinorborn-5-ene-2,3-dicarboximide, piperonyl butoxide, sesamex, and sulfoxide. Piperonyl butoxide interferes with oxidative breakdown of the insecticide that is catalyzed by cytochrome P-450. Piperonyl butoxide and the three other synergists are used to enhance the toxicity of pyrethroids, a group of insecticides that tend to be very expensive and unstable.

Hewlett and Plackett[27] have proposed the following probit model for response of an organism to drug A (applied at dose d_1) plus synergist B (applied at dose d_2):

$$Y = \beta_0 + \beta_1 \log d_1 + \beta_2 \frac{d_2}{\gamma + d_2},$$

where Y is the probit line and β_0, β_1, β_2, and γ are unknown parameters; R[18] or another computer package, such as Stata[28] or Mathematica,[29] can be used to calculate estimates of the parameters; however, Hewlett and Plackett[27] admit that the model has no "definite biological basis." As Dr. Maven has found early in her reading on the problem of mixtures, this statement is applicable to most of the statistical methods published after Bliss's[1] initial work.

A more sensible approach to analyzing the effects of a synergist is to derive a dose–response model for the synergist based on knowledge or a theory about its biological mode of action. For example, Preisler[30] derived a dose–response model to describe the synergistic effect of an enzyme, chitinase, on the biological activity of gypsy moth (*Lymantria dispar* [L.]) nuclear polyhedrosis virus (NPV). The model was based on the mode of action for chitinase suggested by Shapiro et al.,[31] who indicated that the enzyme might disrupt the insects' peritrophic membrane. Once disrupted, greater numbers of viruses can penetrate the midgut and infect susceptible cells. The mode of action of chitinase was modeled as follows.: If $D(t)$ = the amount of NPV reaching susceptible cells by time t, then $D(t)$ is an increasing function of time depending on the amount of NPV present at time zero and the dose of enzyme administered at time zero. Preisler[30] has used the complementary log–log model to estimate the dose–response curve. GLIM was used to calculate values of the parameters.

10.5 CONCLUSIONS

For practical purposes, the simplest hypothesis of pesticide interaction that can be tested experimentally is that of independent, uncorrelated, joint action. This hypothesis means that the effect of one component of the pesticide mixture is not correlated with the effect of the other and is applicable to data from experiments with mixtures consisting of pesticides of different chemical types. If the hypothesis is rejected, two effects are possible. Synergism is the occurrence of significantly greater mortality than predicted by the model, and antagonism is the occurrence of significantly reduced mortality: Another simple, easily tested hypothesis that can be examined is similar (or additive) joint action. This hypothesis is probably applicable when one mixture component is structurally related to the other mixture component. Other hypotheses about pesticide interactions are difficult to test with quantal response bioassays because they are based on statistical rather than biological theories. The effects of synergists are most sensibly modeled by deriving a dose–response model based on a particular biological mode of action.

REFERENCES

1. Bliss, C. I., The toxicity of poisons applied jointly, *Ann. Appl. Biol.* 26, 585, 1939.
2. Hewlett, P. S. and Plackett, R. L., Statistical aspects of the independent joint action of poisons, particularly pesticides. II. Examination of data for agreement with the hypothesis, *Ann. Appl. Biol.* 37, 527, 1950.
3. Hewlett, P. S. and Plackett, R. L., A unified theory for quantal responses to mixtures of drugs: Non-interactive action, *Biometrics* 15, 591, 1959.
4. Hewlett, P. S. and Plackett, R. L., A unified theory for quantal response to mixtures of drugs: Competitive interaction, *Biometrics* 20, 566, 1964.
5. Plackett, R. L. and Hewlett, P. S., Statistical aspects of the independent joint action of poisons, particularly insecticides. I. The toxicity of a mixture of poisons, *Ann. Appl. Biol.* 35, 347, 1948.
6. Plackett, R. L. and Hewlett, P. S., Quantal responses to mixtures of poisons, *J. R. Stat. Soc. B.* 14, 141, 1952.
7. Plackett, R. L. and Hewlett, P. S., A unified theory for quantal responses to mixtures of drugs: The filling to data of certain models for two non-interactive drugs with positive correlation of tolerances, *Biometrics* 19, 517, 1963.

8. Plackett, R. L. and Hewlett, P. S., A comparison of two approaches to the construction of models for quantal responses to mixtures of drugs, *Biometrics* 23, 27, 1967.

9. Ashford, J. R., Quantal responses to mixtures of poisons under conditions of simple similar action: The analysis of uncontrolled data, *Biometrika* 45, 74, 1958.

10. Ashford, J. R., General models for the joint action of mixtures of drugs, *Biometrics* 37, 457, 1981.

11. Ashford, J. R. and Smith, C. S., General models for quantal response to the joint action of mixtures of drugs, *Biometrika* 51, 413, 1964.

12. Ashford, J. R. and Smith, C. S., An alternative system for the classification of mathematical models for quantal responses to mixtures of drugs in biological assay, *Biometrics* 21, 181, 1965.

13. Robertson, J. L. and Smith, K. C., Joint action of pyrethroids with organophosphorous and carbamate insecticides applied to western spruce budworm (Lepidoptera: Tortricidae), *J. Econ. Entomol.* 77, 6, 1984.

14. Robertson, J. L. and Smith, K. C., MIX: A computer program to evaluate interaction between chemicals, *USDA Forest Service Gen. Tech. Rep.*, PSW-112, 1989.

15. Russell, R. M., Robertson, J. L., and Savin, N. E., POLO: A new computer program for probit analysis, *Bull. Entomol. Soc. Am.* 23, 209, 1977.

16. LeOra Software, *POLO-PC: A User's Guide to Probit or Logit Analysis*, 1119 Shattuck Ave., Berkeley, CA, 94707.

17. LeOra Software, *PoloPlus*, LeOra Software, 1007 B St., Petaluma, CA 94952. See http://www.LeOraSoftware.com.

18. R version 1.9.1. Released June 21, 2004. See http://www.r-project.org.

19. LeOra Software, *PoloJR*, LeOra Software LLC, PO Box 562, Parma, MO 63870. See http://www.LeOra-Software.com.

20. LeOra Software, *PoloMix*, LeOra Software, 1007 B St., Petaluma, CA 94952. See http://www.LeOraSoftware.com.

21. LeOra Software, *PoloMixture*, LeOra Software LLC, PO Box 562, Parma, MO 63870. See http://www.LeOra-Software.com.

22. Giltinan, D. M., Capizzi, T. P., and Malani, H., Diagnostic tests for similar action of two compounds, *Appl. Stat.* 37, 39, 1988.

23. Finney, D. J., *Probit Analysis*, Cambridge University Press, Cambridge, England, 1971.

24. Sakai, S., Joint action of insecticides, *Bull. Daito Bunko Univ.* 1, 279, 1969.

25. Busvine, J. R., *A Critical Review of the Techniques for Testing Insecticides*, Commonwealth Agricultural Bureaux, London, 1971.

26. Corbett, J. R., Wright, K., and Baillie, A. C., *The Biochemical Mode of Action of Pesticides*, Academic Press, London, 1984.

27. Hewlett, R. L. and Plackett, P. S., *The Interpretation of Quantal Responses in Biology*, University Park Press, Baltimore, MD, 1979.

28. Stata Corporation, Stata 9 software, 4905 Lakeway Dr., College Station, TX 77845.

29. Wolfram Research *Mathematica* software, Champaign, IL.

30. Preisler, H. K., Analysis of toxicological experiment using a generalized linear model with nested random effects, *Intl. Stat. Rev.* 57, 145, 1989.

31. Shapiro, M., Robertson, J. L., and Preisler, H. K., Enhancement of baculovirus activity on gypsy moth, *J. Econ. Entomol.* 80, 1113, 1987.

CHAPTER **11**

Time as a Variable

Fred Weasley, Paula's neighbor, returns to see her about another problem. It seems that his mother has a question about another pest of tomatoes. Just when the Godfather larvae seem to be under control, her select crop of tomatoes grown and sold for canning is under attack by a subspecies of the Texas leafcutting ant (*Atta texana* [Buckley])—the tomato stem cleaver (*A. texana severalis* [Amputeé]). Once the larvae start to chew on the stem, the tomato drops the fruit 6 hours later. The problem would be minimal if the fruit were ripe, but green tomatoes, while tasty, are not the intended crop.

In truth, our heroine had never really thought about the speed at which a toxicant acted, except in passing. Of course, she has noticed that some pesticides such as the pyrethroids or pyrethrins acted very quickly at low doses. Other kinds of pesticides (e.g., molt inhibitors, juvenile hormone analogues) act comparatively slowly even at high doses. For the sake of the canning tomatoes, Paula and Jessica now examine the matter of time in response of tomato stem cleaver ants to pesticides in more detail.

11.1 PURPOSES OF STUDIES INVOLVING TIME

Time–dose (or time–concentration) relationships to mortality are of practical and theoretical importance in studies of pesticide activity on arthropods. For some species, such as the cone beetle, *Conophthorus ponderosae* Hopkins, the most important criteria for choice of a pesticide are speed of kill and residual activity that persists until the period during which females attack host cones.[1] Female *C. ponderosae* rapidly sever connective tissue in the cone stalk. If cone death is to be prevented, an insecticide must cause death quickly once a beetle contacts the toxicant. In theoretical studies, time trends may also be preliminary indicators of chemical mode of action and detoxification mechanisms.

Data from time–dose–response experiments have frequently been analyzed by modeling time trends separately for each dose[2,3] or by estimating dose trends separately for each time (e.g., Su et al.[4]). But like the use of analysis of variance to test differences in dose–responses of populations, these methods are inefficient because all the data are not used in the estimation procedure. In addition, activity over time is difficult to describe when time and dose trends are considered separately.

Time–dose–response data have also been analyzed by fitting probit lines (with time replacing dose) to mortality data for a fixed dose over time (e.g., Su et al.[4]). Unless different groups are used for each observation period, however, this method is not appropriate for time–response data because responses at different time points are correlated (see Section 11.3.1). Methods for the analysis of correlated data have been described,[5] and a computer program written in the Mathematica[6] language is now available.

11.2 SAMPLING DESIGNS

11.2.1 Alternatives

Bioassays that include both time and dose as variables can have one of two possible designs. With the *independent sampling design*, separate groups of test subjects are treated with a fixed dose. Each group is then observed for a different period of time, and numbers of responses are recorded at each observation period. For the experiment to be valid, responses that occur in a given group of test subjects must be recorded only once.

The other alternative, the *serial sampling design*, involves treatment of each group of subjects with a given dose, treatment with use of several doses for the experiment as a whole, and recording of responses for each dose group at a series of times after treatment. A number of observations are necessary because responses differ with each dose (i.e., a high dose will kill more test subjects in a short period of time, but a low dose will kill fewer subjects during the same interval). Thus, a serial design involves both treatment with a series of observations for each treatment group and treatment with a series of doses and a series of observations with each treatment group.

11.2.2 General Statistical Models

For the independent design, statistical models of time–response data for a fixed dose are straightforward. The probability of response by a given time t can be modeled by the probit, logit, or complementary–log–log (CLL) curves with time (or some function thereof such as the logarithm of time) as the single independent variable. The statistical model for the binary response with time as the explanatory variable is

$$P_i = F(\alpha + \beta t_i), \tag{11.1}$$

where P_i is the probability of response, t_i is the ith time (or the logarithm of time), α is the intercept of the regression line, β is the slope of the regression line, and F is a distribution function (probit, logit, or CLL). The probit and logit models are defined in Chapter 3.

For a fixed exposure time t_j ($j = 1,\ldots, J$), to a chemical at a concentration d_i ($i = 1,\ldots, 1$), probability of mortality of a test subject by time t_j is

$$p_{ij} = 1 - \exp[-\exp(\tau_j + \beta\log_{10}d_i)], \tag{11.2}$$

where $\exp(y) \equiv e^y$, β is an unknown parameter, and τ_j is an unknown categorical variable corresponding with times t_j. The linear part of Equation 11.2, $\tau_j + \beta\log_{10}(d_i)$, is the CLL line (i.e., the linear predictor of the CLL model). The model assumes that $\log_e(-\log_e(1 - p_{ij}))$ is linear in the covariates.

For the independent time–mortality design, τ_j is replaced by the logarithm of t_j and $\log_{10}(d_i)$ is replaced by the constant $\log_{10}(D)$, where D is the fixed dose used in the experiment. This statement of the CLL relationship corresponds with the Weibull function, which has been used to model responses to some chemicals over time.[2]

11.3 ANALYSIS OF INDEPENDENT TIME–MORTALITY DATA

11.3.1 Experimental Design

The experimental design of time–dose–response experiments done with independent sampling is analogous to those dose–response curves described in Chapter 4, in which responses are tallied after a constant time. Instead of time being constant (e.g., mortality tallied after 7 days or some other fixed time), dose is constant and time is varied.

At first, Dr. Maven is baffled how to use time as a factor, until she realizes that in a dose–response bioassay, test groups are treated with *different* doses or concentrations. In time experiments at a constant dose, responses of *different treatment groups* are recorded at each observation period. Once the experimental dose or concentration has been selected, preliminary experiments are necessary to bracket the range of times (five to seven) at which 5% to 95% mortality occurs. At least three replications with all of the time groups should then be done. At least 240 test subjects should be tested in the complete experiment.

As with dose–response experiments, selection of the optimal times depends on the lethal time (LT) of interest and the number of times used. The information in Chapter 5 (Table 5.1) can be used for selection of the optimal experimental design (placement of times, numbers of time points, number of individual times tested) by substituting LT for effective dose throughout. The first column (number of doses) should be replaced with the heading "Number of Times."

11.3.2 Limitations and Constraints

For a valid experiment with an independent design, a different group of test subjects must be used for assessment of response at each time. If the same group of test subjects is observed at several time intervals, responses at different time points will be correlated. Analysis of correlated response data requires the use of distinctive computer programs that do multinomial analysis. Responses at different time points violates the premise of independent binomial response assumed in POLO,[7] GLIM,[8] PoloPlus,[9] PoloSuite,[10] or SAS.[11] One such special program specifically for correlated data[5] is written in Mathematica[6] language. Unfortunately, many published reports[5,8] describe analyses of data collected by sequentially recording the mortality within the same group of individuals, followed by probit or logit analysis with time as the independent variable. Results of such analyses are not valid.

A final problem must be kept in mind when time or the logarithm of time is used as the independent variable in probit, logit, or CLL models. If the function of time used in the regression line increases indefinitely as time increases, then the probabilities of mortality will increase to 1. Such a model suggests that all test subjects will eventually die from the treatment. Not true. Many insects will die from causes other than the pesticide. Therefore, the probability of mortality will reach a plateau that is usually <1. As a result, a probit, logit, or CLL model with time (or logarithm of time) as the independent variable will not be appropriate for these data except at higher doses.

The problem of plateaus of <100% mortality (i.e., maximum $P \leq 1$) is especially relevant to studies of pesticide resistance in which LTs are used to compare population responses (e.g., Cochran[12]).[13] A dose selected on the basis of responses of susceptible populations will probably not kill all members of resistant populations regardless of the observation times. A dose that will affect all members of the most resistant population might be used instead, but the observation times for each susceptible population should be chosen carefully so that most test subjects are not dead when the first count is made. Careful preliminary experiments will be necessary to select the times that are appropriate for each population; use of OptiDose[14] can facilitate their selection to obtain the most precise estimate of the LT of interest.

11.4 ANALYSIS OF SERIAL TIME–MORTALITY DATA

For Paula Maven's problem with tomato stem cleaver ants on tomatoes, use of the serial time–mortality design is the ideal solution for her problem. This design, in which the same test subjects are followed over time, is the most economical way to use test subjects (not that Jessica's supply of ants is running out). Another advantage of using the serial design is that Dr. Maven and Jessica can gain more insight into the mode of action of a particular pesticide and the effect of dose on speed of lethal action. Finally, a model with a nonparametric (categorical) function of time can be fitted when the data for all doses are analyzed simultaneously. Such a model is more general than one with time or logarithm of time as the independent variable (e.g., Equation 11.1) and it includes models with plateaus with probability values <1.

11.4.1 Experimental Design

For this kind of experiment, application rates should be selected as described in Section 2.2. Times at which mortalities are tallied should be chosen so that 5% to 95% of the total mortality that occurs with each dose is included in the observations. In other words, if 20% of your control insects die in 48 hours, do not design an experiment that lasts 7 days.

At each observation period when mortality is recorded, each dead test subject should be removed from the treatment container, held in a separate container, and its condition verified at the time of the next observation. This procedure ensures that a test subject is counted as dead at only one observation interval and that a subject categorized as dead is, in fact, dead. If moribund individuals will be included with dead individuals, the criteria for moribund should be clear prior to starting the observations.

Unless other variables such as body weight are intentionally included in an experiment, they must be controlled. For example, a narrow range of body weights (average or individual) can be used to exclude body weight as a variable. An alternative procedure is to record all body weights and test for the significance of body weight as a variable in the time–dose–response regression. Likewise, for Lepidoptera, use of specific and identical developmental stage or groups of stages (e.g., neonates or mature larvae) can help minimize development as a variable. Measurement of head capsule width for a particular species is a good way to determine instar stage.

For the experiments, Dr. Maven decides to have Jessica test commercial formulations of two pesticides (pyrethrins and acephate) that she observed acted quickly on the Godfather larvae in preliminary experiments. But she decides that the only way to get results that might predict a response when the pesticide is applied is to use a bioassay that will simulate natural infestation. Jessica places 10 adults of each sex (pupa are sexed, 4 abdominal rings are male, 5 are female) into screened containers with a tomato plant and waits for oviposition. Three days later, she collects eggs from the tomato plant and gently places 10 eggs on a fresh tomato plant. Each tomato plant is in first phase fruit development (~49 DAP) and bears at least 10 green fruits. Dr. Maven uses first instars because the results of other experiments suggest that this is the most susceptible developmental stage. But even if the first instars were not most susceptible, she would have used them as the target for control to minimize damage to the tomato crop. The caterpillars, growing by the second, are left undisturbed for 24 hours before the spray application begins. Each plant is sprayed with one of five concentrations of each chemical using a Crown Spray tool sprayer that evenly coats each leaf without the need for a surfactant. Two plants are sprayed in a replication with each concentration, and each experiment is replicated three times for each sex. Controls for each pesticide are sprayed with water only; a control group is included with each replication with each chemical. All concentrations of a pesticide are sprayed in order of lowest to highest in each replication, and both pesticides are tested in a replication.

With help from high school students, Dr. Maven and Jessica obtain mortality counts at 4, 8, 12, 24, 48, 72, 96, 120, 144, and 168 hours after treatment. Because changes in observers (two high school sophomores quit after 1 week to bag groceries) introduce another possible variable into the experiment. Dr. Maven holds a group session about how to tell whether a larva is dead. Insects were to be touched by a blunt probe and their movement is observed. A larvae was considered to be alive if it was about to walk or move its head or legs when prodded. Larvae that exhibited no walking and only twitching of the abdomen were considered to be moribund and combined with dead insects (no movement) for analysis.

11.4.2 Statistical Methods

With the serial design, standard techniques that are suitable for the independent design are *not* applicable for data analyses because responses of test subjects over time are not independent. Thus, analyses such as those described by Su et al.[4] and Cochran[12] are not appropriate: When counts are made on the same individuals at each time period, standardized probit or logit techniques with time as the independent variable should not be used. For a given dose, the numbers of responses at each time out of the total number treated with that dose have a multinomial distribution. Therefore, analyses of dose–mortality data over time can be done with the procedure described in Chapter 14.

The conditional probability that a test subject responds by time t_k given its failure to respond at time t_{k-1} is modeled by

$$y_{ijk} = \gamma_{jk} + \beta_j \log_{10}(d_i + d_0), \tag{11.3}$$

where y_{ijk} is the probit, logit, or CLL line for test subjects in group j treated with dose d_i at the beginning of the experiment (t_0). If group j includes separate subgroups of males and females, then $j = 1, 2$. $\gamma_{j,k}$ is the intercept (one for each sex and time interval), β_j is the slope (one for each sex), and $d_0 = 10^a$. The value a is defined as

$$a = \log(d_0) = \log(d_i) - \frac{d_1}{d_2 - d_1} [\log(d_2) - \log(d_i)], \tag{11.4}$$

where $0 < d_1 < d_2$. Equation 11.4 provides a means to deal with a dose of zero.

The CLL model is ideal for the analysis of time–mortality data with a serial design. In an extensive series of experiments with western spruce budworm, Preisler and Robertson[15] have shown that the cumulative probability of a subject responding by time t_k can be adequately approximated by the CLL model. The line for cumulative probability of response by time t_k is given by

$$y_{ijk} = \tau_{jk} + \beta_j \log_{10}(d_i + d_0), \tag{11.5}$$

where β_j and d_0 are the same as in Equation 11.3 and where

$$\tau_{jk} = \log_e \left(e^{\gamma_{j1}} + e^{\gamma_{j2}} + \cdots + e^{\gamma_{jk}} \right). \tag{11.6}$$

The relationship between γ and τ in Equation 11.6 does not hold for the standard probit or logit models.

As with polytomous response data, conditional probabilities given in Equation 11.5 can be used to calculate maximum likelihood (ML) estimates of the parameters to test hypotheses of parallelism or equality. This estimation procedure is appropriate because conditional responses are independent binomial variables. Therefore, standard programs for estimation of parameters of probit, logit, or CLL lines can be used to obtain the ML estimates. However, cumulative probabilities in Equation 11.5 are needed to calculate parameters of biological interest, such as estimates of lethal dose over time.

11.4.3 Estimation

For estimation of Equation 11.5, two data sets are needed: (1) the number of subjects (r_{ijk}) in category ij (i.e., of group j that have been treated with dose d_i) that died in time interval (t_{k-1}, t_k) and (2) the number of subjects (n_{ijk}) in category ij that were still alive at time t_{k-1}. The latter variable therefore equals the total number of insects in category ij minus all of the insects that died by time t_{k-1}.

Because the number of responses r_{ijk} out of n_{ijk} are independent binomial variables, a software package that performs binomial regressions with the CLL line (e.g., R[16] or GLIM[8]) can be used to estimate the parameters in Equation 11.4 and to compute maximum values of likelihood functions. The significance of categories on the CLL line can then be tested with likelihood ratio (LR) tests.

11.4.3.1 Estimation of Response Probabilities

For estimation of response probabilities, the model with the least number of parameters that fits the data adequately is used to calculate estimates of the probabilities of response (P_{ijk}) up to time t_k. For example, if the effect of sex on the responses in all time intervals is not significant, then the model with equal lines for both sexes can be used. The CLL model becomes

$$\hat{y}_{ijk} = \hat{\gamma} + \beta \log_{10}(d_i + d_0) \tag{11.7}$$

$$\hat{\tau}_k = \log_e \left(e^{\hat{\gamma}_1} + e^{\hat{\gamma}_2} + \cdots + e^{\hat{\gamma}_k} \right) \tag{11.8}$$

$$\hat{P}_{ijk} = 1 - \exp\left\{ -\exp\left[\hat{\tau}_k + \hat{\beta} \log_{10}(d_i + d_0) \right] \right\}. \tag{11.9}$$

Plots of $\hat{p}_{ijk}$ against time can also be used to study the effect of dose on speed of lethal action (see Section 11.4.3.3).

11.4.3.2 Estimation of Lethal Doses over Time

The parameter $LD_x(t_k)$ is the dose that causes $x\%$ mortality by time t_k. An estimate of $LD_x(t_k)$ is $10^{\hat{\theta}_k}$, where

$$\hat{\theta}_k = \frac{a_x - \hat{\tau}_k}{\hat{\beta}}$$

and

$$a_x = \log_e[-\log_e(1 - x/100)]. \tag{11.10}$$

Based on the asymptotic approximation of variances of functions of random variables, the equation for the variance of $\hat{\theta}$ is

$$\text{var}\left(\hat{\theta}_k\right) = \frac{1}{\hat{\beta}^2}\left[\hat{\theta}_k^2 \text{ var}\left(\hat{\beta}\right) + \text{var}\left(\hat{\tau}_k\right) + 2\hat{\theta}_k \text{ cov}\left(\hat{\tau}_k,\hat{\beta}\right)\right] \tag{11.11}$$

$$\text{cov}\left(\hat{\tau}_k,\hat{\beta}\right) = \frac{1}{e^{\hat{\tau}_k}}\sum_{i=1}^{k}e^{\hat{\gamma}_i}\text{ cov}\left(\hat{\beta},\hat{\gamma}_k\right) \tag{11.12}$$

$$\text{var}\left(\hat{\tau}_k\right) = \frac{1}{e^{2\hat{\tau}_k}}\sum_{i=1}^{k}\sum_{i'}^{k}e^{\hat{\gamma}_i+\hat{\gamma}_i}\text{ cov}\left(\hat{\gamma}_i,\hat{\gamma}_{i'}\right). \tag{11.13}$$

Finally, estimates for variances and covariances of the estimates $\hat{\beta}$ and $\hat{\gamma}_k$ are obtained from the output of the statistical program such as GLIM8 that was used to fit the CLL model to Equation 11.3. Plots of $LD_x(t^k)$ versus time permit comparisons of effectiveness to be made at selected times after the test subjects are first exposed to different chemicals.

11.4.3.3 Example

Some of the data gathered by Dr. Maven and Jessica are shown in Table 11.1. Dr. Maven uses CLL regression to estimate the conditional binomial model. The GLIM[8] output begins with a listing of the data (part of which is shown in Figure 11.1a). Under the column labeled "Sex"; 1 designates males and 2 designates females. "Time" is the time elapsed from the beginning of the experiment until the time an observation was made. "Dose" is self-explanatory, and "Response" is the number of tomato stem cleaver ants that died since the last time an observation was made. Finally, "Conditional Response" is the number of insects, exposed to a given concentration, that were still alive when the time interval began. For example, in Figure 11.1, line 4, the conditional total of 58 equals the conditional total in line 3 minus the 2 insects that died in the time interval.

Figure 11.1b shows the commands for testing the hypotheses of parallelism and equality of CLL lines for male and female ants. Because control mortality (dose 0) was zero, d_0 is set to zero. The output from these commands (Figure 11.1c) shows the information necessary for hypothesis testing. H_1 is the hypothesis that the CLL lines

Table 11.1 Number of Responses[a] of Tomato Stem Cleaver Males or Females in Each Time Interval after Application of Acephate to Tomato Plants

Sex	Dose	Time Interval in Hours after Treatment									
		4	8	12	24	48	72	96	120	144	168
Males	0	0	0	0	0	0	0	0	0	0	0
	1	3	11	5	5	0	2	1	0	4	5
	2	11	14	5	7	3	3	1	0	6	2
	3	12	17	7	4	3	6	2	0	3	1
	5	13	12	15	7	5	2	0	3	1	2
	10	13	17	9	16	5	0	0	0	0	0
Females	0	0	0	0	0	0	0	0	0	0	0
	1	2	13	10	5	1	2	0	0	2	5
	2	16	15	9	6	1	0	1	0	4	0
	3	12	20	8	7	3	2	0	0	2	0
	5	11	15	16	6	5	1	0	1	0	2
	10	15	23	8	12	2	0	0	0	0	0

[a] The total number of test subjects at each dose is 60.

for males and females are parallel; statistics for this test are given in lines 5 and 9. The difference in the values of the LR test statistic is 2.77 (i.e., residual deviance = 130.32; line 9) minus 127.55 (i.e., residual deviance = 127.55; line 5). Degrees of freedom for the test are 78 (line 9) minus 77 (line 5), or 1. The P value for 2.77 with 1 degree of freedom (df) is >5%, 0.05, so H_1 cannot be rejected. H_2 is the hypothesis that the CLL lines for males and females are equal. The LR test statistic is 147.53 (line 13) minus 127.55 (line 5). This value, 19.98, is greater than the critical χ^2 value for 11 df (88 [line 13] − 77 [line 4]). Therefore, the hypothesis of equality is rejected, and the parallel sex lines model is assumed to be the best.

Because the intercepts for males and females are significantly different, parameters of the conditional binomial model with equal slopes and separate intercepts for both sexes are computed next (see Figure 11.1b). In Figure 11.1c (the output from running Figure 11.1b), the intercept for males at a given interval is computed by adding the "SEX1" estimate to its corresponding time interval estimate. For example, the intercept for a male at time interval 2 (8 hours) is (−2.1834 [line 19] + 0.5740 [line 21] = −1.6094). The intercept for females at a given time interval is computed by adding the "SEX2" estimate to its corresponding time interval estimate, as well as the corresponding interaction estimate between "SEX2" and the time interval. For example, the intercept for a female at time interval 2 is (−2.0675 [line 20] + 0.5740 [line 21] + 0.2572 [line 30] = −1.3522). Recall that in the best model, there was no significant interaction between sex and dose (i.e., slopes are parallel); thus, the rest of the equation for a point estimate consists of adding one more term, which is the log.dose (slope) estimate times the actual log.dose. For example, the point estimate for a male at time interval 2 with log.dose = 2 is (−2.1834 [line 19] + 0.5740 [line 21] + (1.0732 [line 18] × 2)) = 0.537.

!This is the acephate data file
!The columns are sex, time (hours), dose, response, conditional total

	Sex	Hours	Dose	Response	Conditional_Response
1	1	4	1	2	60
2	1	8	1	10	58
3	1	12	1	4	48
4	1	24	1	4	44
5	1	48	1	0	40
6	1	72	1	1	40
7	1	96	1	2	39
8	1	120	1	0	37
9	1	144	1	3	37
10	1	168	1	4	34
11	1	4	2	10	60
12	1	8	2	12	50
13	1	12	2	4	38
14	1	24	2	6	34
15	1	48	2	1	28
16	1	72	2	2	27
17	1	96	2	1	25
18	1	120	2	1	24
19	1	144	2	5	23
20	1	168	2	3	18
21	1	4	3	10	60
22	1	8	3	15	50
23	1	12	3	6	35
24	1	24	3	2	29
25	1	48	3	4	27
26	1	72	3	2	23

(a)

Figure 11.1 Complementary-Log-Log (CLL) regression of Time-Dose-Mortality data done with the R Program. (*Continued*)

What we are really interested in calculating is the probability of death by a certain time frame at a certain dose. We show how this can be done through use of Equation 11.6 for a male at time interval 2 (8 hours) and dose 2,

$$\hat{\tau}_2 = \log_e(e^{-2.1834} + e^{-2.1834+5.74}) = -1.163$$

From Equation 11.7,

$$p_{22} = 1 - \exp\{-\exp[-1.163 + 1.0732\log_{10}(2)]\} = 0.351$$

Thus, a tomato leaf cleaver ant has an estimated 35.1% chance of being dead 8 hours after treatment (i.e., sprayed) with 2 g/ha of acephate.

$C this program tests the hypotheses of parallelism and equality
$C of CLL-lines for males and females.
$Units 100$ data sex time D R N$dinput 31$

```
1    setwd('C:/Users/Brad/Documents/Entomology/data/')
2    DATA=read.csv('DATAch11new.csv')
3    colnames (DATA)=c('SEX', 'HOURS', 'DOSE', 'RESPONSE', 'CONDITIONAL_RESPONSE')
4    attach(DATA,2)
5    log.dose=log10(DOSE)
6    HOURS=as.factor (HOURS)
7    SEX=as.factor (SEX)
8    r=ifelse(RESPONSE==0 & CONDITIONAL_RESPONSE==0,
9              0,RESPONSE/CONDITIONAL_RESPONSE)
10
11   #separate sex lines
12   modl=glm(r~SEX*HOURS+ SEX*log.dose-1,binomial (link='logit'),
13   .             weight=CONDITIONAL_RESPONSE)
14   het=mod1$dev/mod1$df.residual
15   print(paste('HETEROGENEITY FACTOR =',round(het,4)))
16   summary(mod1,dispersion=het,cor+f)
17
18   #Parallel sex lines
19   mod2=glm(r~log.dose+SEX*HOURS-`,binomial(link='logit'),
20              weight=CONDITIONAL_RESPONSE)
21   summary(mod2,dispersion=het,cor=F)
22
23   #Equal sex lines
24   mod3=glm(r~HOURS + log.dose - 1,binomial (link+'logit'),
25              weight=CONDITIONAL_RESPONSE)
26   summary(mod3,dispersion=het,cor=F)
```

(b)

Figure 11.1 (Continued)   Complementary-Log-Log (CLL) regression of Time-Dose-Mortality data done with the R Program.   (*Continued*)

Figure 11.2a and b shows plots of the probabilities for all time points between 0 and 168 hours after treatment, for all application rates of acephate for males (Figure 11.2a) and females (Figure 11.2b). As the acephate dosage increases, probabilities of mortality also increase regardless of sex.

But the time required to reach the plateau when all insects will die appears to different for males and females. It appears that females die more rapidly and abundantly than males do at a given dose. Based on these results, Dr. Maven recommends that Fred Weasley apply pyrethrins to his crop of tomatoes because the highest rate kills almost all of the *Atta texana severalis* leaf cleaver ants within 30 minutes. *I am unsure what to recommend to Dr. Maven in this case.*

Table 11.2a and b list the estimates of LD_{50} and LD_{90} for *A. texana severalis* (with their 95% confidence intervals) for tomato leaf cleaver ants treated with acephate

```
1    [1] "HETEROGENEITY FACTOR = 1.6565"
2
3    #separate sex lines
4    Null deviance: 1475.63 on 99 degrees of freedom
5    Residual deviance: 127.55 on 77 degrees of freedom
6
7    #parallel sex lines
8    Null deviance: 1475.63 on 99 degrees of freedom
9    Residual deviance: 130.32 on 78 degrees of freedom
10 .
11   #equal sex lines
12   Null deviance: 1475.63 on 99 degrees of freedom
13   Residual deviance: 147.53 on 88 degrees of freedom
14
15.  #best model (parallel sex lines)
16   Coefficients:
```

17 z.	Estimate	Std. Error	z value	PR($>|z|$)	
18 log. Dose	1.07323	0.2009	5.342	9.19e-08	***
19 SEX1	-2.1834	0.2540	-8.595	< 2e-16	***
20 SEX2	-2.0675	0.2255	-9.168	<2e-16	***
21 HOURS8	0.5740	0.2872	1.999	0.0457	*
22 HOURS12	0.2078	0.3248	0.640	0.5222	
23 HOURS24	0.4064	0.3369	1.206	0.2278	
24 HOURS48	-0.2576	0.4296	-0.600	0.5488	
25 HOURS72	-0.6894	0.5250	-1.313	0.1891	
26 HOURS96	-0.9026	0.5884	-1.534	0.1250	
27 HOURS120	-1.2674	0.6965	-1.820	0.0688	
28 HOURS144	0.1851	0.4402	0.420	0.6741	
29 HOURS168	-0.2562	0.5347	-0.479	0.6319	
30 SEX2:HOURS8	0.2572	0.3888	0.662	0.5082	
31 SEX2:HOURS12	0.4761	0.4390	1.084	0.2782	
32 SEX2:HOURS24	0.3411	0.4691	0.727	0.4671	
33 SEX2:HOURS48	0.2473	0.6124	0.404	0.6864	
34 SEX2:HOURS72	-0.4293	0.8689	-0.494	0.6213	
35 SEX2:HOURS96	-1.5968	1.4383	-1.110	0.2669	
36 SEX2:HOURS120	-1.2176	1.4860	-0.819	0.4126	
37 SEX2:HOURS144	-0.5820	0.7068	-0.823	0.4103	
38 SEX2:HOURS168	-0.2011	0.7976	-0.252	0.8010	

(c)

Figure 11.1 (Continued)   Complementary-Log-Log (CLL) regression of Time-Dose-Mortality data done with the R Program.

and for increasing times of exposure for males and females, respectively. These values were calculated using Equations 11.10–11.13. For example, the LD_{50} for a male after 8 hours of exposure is calculated from Equation 11.10, with $k = 2$ and $x = 50\%$ as

$$a_{50} = \log_e[-\log_e(1-0.5)] = -0.367$$

$$\hat{\theta}_2 = \frac{-0.367-(-1.163)}{1.0732} = 0.742 \text{ and } LD_{50}(8) = 0.742 = 5.517.$$

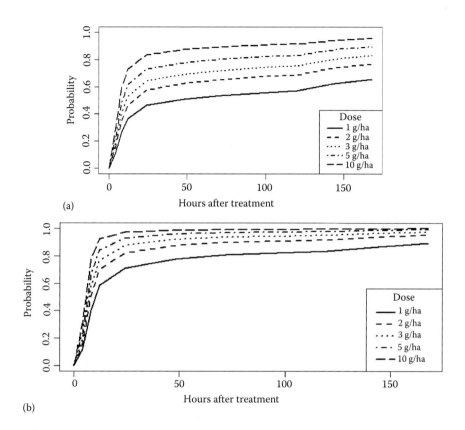

Figure 11.2 Probabilities of death over time for pesticides for which the amount applied has no effect on speed of lethal action for (a) males and (b) females.

Table 11.2 Estimates of LD_{50} and LD_{90} for Male and Female Texas Stem Cleaver Ants on Tomato Plants with Increasing Times of Exposure to Acephate

LD	Hours after Exposure						
	4	8	12	24	48	72	96
a. Male							
50	49.26	5.52	2.51	1.27	0.96	0.81	0.72
90	648.02	72.51	32.99	16.66	12.57	10.66	9.41
b. Female							
50	38.42	1.89	0.61	0.28	0.19	0.16	0.14
90	505.36	24.89	7.99	3.76	2.50	2.05	2.05

11.5 CONCLUSIONS

Bioassays involving both time and dose as independent variables can have one of two possible designs. Separate groups of test subjects are treated with a given dose when the independent sampling design is used. Each group is observed for a different period of time, when responses are tallied. Counts cannot be made on the same individuals at each time period. With the serial sampling design, subjects are treated with a series of doses, and responses for each dose are recorded at different times. Of the two designs, the serial sampling design is more efficient.

REFERENCES

1. Haverty, M. I. and Wood, J. R., Residual toxicity of eleven insecticides to the mountain pine cone beetle, *Conophthorus monticolae* Hopkins, *J. Ga. Entomol. Soc.* 16, 77, 1981.
2. Hewlett, P. S., Time from dosage to death in beetles, *Tribolium castaneum*, treated with pyrethrins or DDT, and its bearing on dose-mortality relations, *J. Stored Prod. Res.* 10, 27, 1974.
3. Shapiro, M., Preisler, H. K., and Robertson, J. L., Enhancement of nucleopolyhedrosis virus activity on gypsy moth (Lepidoptera: Lymantriidae) by chitinase, *J. Econ. Entomol.* 80, 1113, 1987.
4. Su, N-Y., Tamashiro, M., and Haverty, M. I., Characterization of slow-acting insecticides for remedial control of the Formosan subterranean termite (Isoptera: Rhinotermitidae), *J. Econ. Entomol.* 80, 1, 1987.
5. Throne, J. E., Weaver, D. K., Chew, V., and Baker, J. E., Probit analysis of correlated data: Multiple observations over time at one concentration, *J. Econ. Entomol.* 88, 1510, 1995.
6. Wolfram, S., *Mathematica*, Wolfram Research Champaign, IL. 2005.
7. Russell, R. M., Robertson, J. L., and Savin, N. E., POLO: A new computer program for probit analysis, *Bull. Entomol. Soc. Am.* 23(3), 209–213, 1977.
8. Payne, C. D., ed., *The GLIM System Release 3.77 Manual*, Numerical Algorithms Group, Oxford, England, 1978.
9. LeOra Software, PoloPlus, LeOra Software, 1007 B St., Petaluma, CA 94952, 2003. See http://www.LeOraSoftware.com.
10. LeOra Software, PoloSuite, LeOra Software, PO Box 562, Parma, MO 63870, 2016. See http://www.LeOra-Software.com.
11. SAS Institute, http://www.sas.com/technologiest/anaytics/statistics/stat.
12. Cochran, D. G., Selection for pyrethroid resistance in the German cockroach (Dictyoptera: Blatellidae), *J. Econ. Entomol.* 80, 1117, 1987.
13. Preisler, H. K., Hoy, M. A., and Robertson, J. L., Statistical analyses of modes of inheritance for pesticide resistance, *J. Econ. Entomol.* 83, 1649, 1990.
14. LeOra Software, LLC. 2016. OptiDose. See http://www.LeOra-Software.com.
15. Preisler, H. K. and Robertson, J. L., Analysis of time-dose-mortality data, *J. Econ. Entomol.* 83, 1990.
16. R version 1.9.1. Released June 21, 2004. See http://www.r-project.org.

CHAPTER 12

Binary Quantal Response with Multiple Explanatory Variables

Jessica has been rearing a laboratory colony of *Patronius giganticus* for 18 months, and the moths have sustained a few population crashes as they transitioned from natural to artificial diet. Her boss, Dr. Maven, assures Jessica that she has done nothing wrong but suggests she split the colony between those fed artificial diet and those fed a natural leaf diet to reduce the chance of complete colony collapse. They now have two susceptible populations of *P. giganticus* reared on two mediums. How will they compare their data to field populations?

A protocol change is in order, so she brings the matter to Paula's attention. Dr. Maven ponders this situation—what should they do? Jessica could conduct another set of bioassays for insects reared on artificial diet, but what if the artificial diet affects the pesticide efficacy? If it does, all of the previous results will have lost their value for comparative purposes. Once again, an online literature search for similar research reveals that multiple probit or logit regression might be the most efficient method to handle this burdensome problem. Just in case diet is a significant factor, Jessica will record binary responses (dead or alive), doses, and diet for each larva. This type of bioassay is a binary quantal response experiment with multiple explanatory variables.

12.1 EARLY EXAMPLES AND INEFFICIENT ALTERNATIVES

Examples of the use of multiple regression techniques in bioassays with arthropods were scarce in the published biological literature before publication of the first edition[1] of this book. Finney[2] confined his discussion of multiple probit analyses to a discussion with two quantitative variables. In the particular situation he described ("probit plane"), the logarithm of each variable is linearly related to the probit of response when the other variable is fixed. His example includes detailed instructions for manual calculations, but it would be much more efficient if a computer program was available. In the first edition of this book, the authors described computer programs and their output and interpretation. In the second edition, a more user-friendly software was developed, PoloEncore,[3] that could examine multivariables. However, this software required a lengthy input file to be created that was better utilized by a statistician than an entomologist. For the updated version of the software, PoloMulti,[4] it is easier to enter data and interpret results, but the original equations and inner workings remain the same.

Perhaps to avoid such complicated calculations and confusing computer outputs, many biological scientists have chosen (and continue to choose) an alternative that is primitive compared with estimation by multiple regression. Rather than using multiple regression techniques, these investigators have done multiple experiments (with univariate analyses) in attempts to show the effects (but not necessarily test the significance) of variables other than dose on response. For example, early research on the role of insect body weight on response was done by testing separate groups of selected weights, estimating their responses by simple (univariate) probit analysis, and making comparisons at LD_{50} or LD_{90} as the basis for concluding that weight did or did not significantly affect response.[5–8]

The primary reason for use of binary response bioassays with multiple explanatory variables is simple: Compared with a single experiment in which significance of several variables is tested simultaneously, multiple experiments with single variables are inefficient.

12.2 GENERAL STATISTICAL MODEL

The general model for solution of a multiple probit or logit problem is $y = \beta_1 x_1 + \beta_k x_k$, where y is the probit or logit of response and x_1 through x_k are the explanatory variables including the constant, $x_1 = 1$. Regression coefficients are β_1 through β_k for constant term and explanatory variables. The presence of more than one explanatory variable makes the statistical methods used to estimate the probability of response and values of the coefficients far more complex than those previously described for binary quantal response with only one explanatory variable.

As in any binary response model, an individual can respond in one of two ways— dead or alive. The estimated probability p that an individual is dead is $F(\beta'x_i)$, where F is a distribution function, $\beta' = (\beta_1, \ldots, \beta_k$ is a $K \times 1$ vector of unknown regression coefficients, and $x_1 = (x_{li}, \ldots x_{ki})$ is a $K \times 1$ vector of explanatory variables (including a constant) for individual i. For the probit model, $p_i = F(\beta'x_1 = \phi(\beta'x_1)$, where ϕ is the standard normal distribution function. For the logit model,

$$ p_i = F(\beta'x_1 = \frac{1}{1+e^{-\beta'x_i}}. $$

The maximum likelihood method can be used to estimate either model. Details of the statistical methods for these estimation procedures are given by Finney,[2] Domencich and McFadden,[9] Russell et al.,[10] Maddala,[11] and LeOra Software LLC.[12]

12.3 TYPES OF VARIABLES IN MULTIPLE REGRESSION MODELS

As specified by the multiple regression model, the linear predictor (e.g., bit or logit plane [two covariates] or hyperplane [three or more covariates]) is modeled with variates in addition to dose. Multiple regression models are appropriate for many kinds of bioassays. For example, the effects of continuous variables such as body weight and temperature on response can be estimated. When continuous covariates of dose are included in the model, variability in response usually decreases. The subject of body weight as a variable will be discussed in more detail in the next chapter (Chapter 13).

Multiple regression models can also be used to study differences in response among groups of experiments by the addition of categorical covariates (e.g., sex, source of test subjects, types of diets) to the linear predictor. Examples in this chapter demonstrate analyses with such variables. Finally, as discussed in Chapter 10, use of multiple regression models is appropriate in the study of the joint action of pesticide mixtures. In experiments with mixtures, the multiple covariates are doses of the various pesticides in the mixture.

12.4 COMPUTER PROGRAMS

Several computer programs are available for multiple probit or logit analyses of data from pesticide bioassays. Two programs, PoloEncore and PoloMulti, were written specifically for studies including up to nine explanatory variables (especially body weight) on response of insects to pesticides. Reichenbach and Collins[13] used two programs in the UCLA Biomedical Computer Program P Series[14] to examine the effects of temperature and humidity on the toxicity of propoxur to German cockroaches and western spruce budworm. GLIM[15] was used to estimate multiple probit or logit regressions, as illustrated by the analyses used by Shapiro et al.,[16] to show that an enzyme, chitinase, significantly enhanced the activity of a baculovirus ingested by gypsy moth, *Lymantria dispar* (L.), larvae.

12.5 MULTIPLE PROBIT ANALYSIS: EXAMPLE FROM POLOMULTI

A new computer program, PoloMulti,[4] is shown here for the first time. This program is the Microsoft Windows- and Apple OS-compatible version of a previous Polo program, POLOEncore,[3] that was illustrated in the second edition[17] of this book. This program should facilitate use of multiple regression techniques for results of bioassay on arthropods because tedious input and convoluted output have been removed.

To solve the complex analyses with the larvae of exponential weight gain, Dr. Maven begins her education in multiple regression methods by working through a sample problem. Among the examples of multiple probit analyses given in the PoloMulti online manual, data from experiments done by Tattersfield and Potter[18] are used to illustrate a relatively simple analysis with two covariates. The same problem—parallel planes—was also described in detail by both Finney[2] and Busvine.[19]

Tattersfield and Potter exposed red flour beetles, *Tribolium castaneum* (Herbst), to pyrethrum either as a direct spray or as a film deposited on glass (Table 12.1). The response variable for each bioassay is binary (dead or alive); the explanatory variables are

Table 12.1 Tattersfield and Potter: Beetles Exposed to Pyrethrum
Either as Spray or Deposit

Plane	Pyrethrum Concentration	Weight of Deposit	Exposed	Responses
1	5	2.9	27	1
1	10	2.9	29	15
1	20	2.9	30	27
1	40	2.9	28	28
1	5	5.7	29	4
1	10	5.7	29	19
1	20	5.7	27	26
1	40	5.7	30	30
1	5	10.8	30	6
1	10	10.8	24	15
1	20	10.8	31	31
1	40	10.8	19	19
2	5	2.9	29	3
2	10	2.9	30	10
2	20	2.9	29	24
2	40	2.9	29	29
2	5	5.7	27	4
2	10	5.7	28	14
2	20	5.7	28	27
2	40	5.7	29	29
2	5	10.8	28	8
2	10	10.8	28	17
2	20	10.8	28	26
2	40	10.8	17	17

Source: Tattersfield F, Potter C, Ann. Appl. Biol., 30, 259, 1943.[18]

concentration of pyrethrum and spray deposit (i.e., the amount of the carrier solvent). If both explanatory variables are significant, then the response will be predicted by a plane or line. The probit or logit of response might be linearly related to the measurement (or logarithm of the measurement) of each factor when the other factor is held constant.

12.5.1 Statistical Model

The probit model that expresses the effect of pyrethrum spray is

$$y_s = \alpha_s + \beta_{1s} x_1 + \beta_{2s} x_2, \tag{12.1}$$

where y_s is the probit of percent mortality, x_1 is the pesticide concentration in mg/ml (a continuous variable), and x_2 is the spray deposit in mg/cm^2 (another continuous variable). The regression coefficients are α_s for the constant, β_{1s} for spray concentration, and β_{2s} for deposit weight.

For the lethal effect of pyrethrum film, the model is

$$y_f = \alpha_f + \beta_{1f} x_1 + \beta_{2f} x_2, \tag{12.2}$$

where y_f is the probit of percent mortality, x_1 is concentration, and x_2 is weight of the deposit on the glass. The regression coefficients are α_f for the constant, β_{1f} for concentration, and β_{2f} for weight of the deposit or film.

12.5.2 Hypotheses Tests

Likelihood ratio (LR) tests can be used to test two hypotheses about these data. These hypotheses are, first, that the planes (i.e., dose–response curve lines) for the spray and residual film are parallel; second, that the planes for the spray and residual film are parallel and equal; and third, the planes are equal but not necessarily equal. If you wish to test whether or not two planes or lines are statistically alike or different, the second hypothesis is the one that is tested.

The LR test compares two values of the logarithm of the likelihood function. The first, $L(\omega)$, is the value when the log likelihood is maximized with no restrictions. The second, $L(\omega)$, is the maximum of the likelihood function subject to the restrictions of the hypothesis being tested.

The hypothesis of parallel planes is H:(P): $\beta_{1s} = \beta_{1f}$, $\beta_{2s} = \beta_{2f}$. L_s and L_f are the respective values of the maximum of the log likelihood for Equations 12.1 and 12.2. The value of $L(\Omega)$ is the sum of L_s and L_f. The model with the restrictions imposed is

$$y = \alpha_s x_s + \alpha_f x_f + \beta_1 x_1 + \beta_2 x_2. \tag{12.3}$$

Categorical variables are used to solve this equation, so that $x_s = 1$ for spray; $x_f = 1$ for film; $x_s = 0$ for film; and $x_f = 0$ for spray. The value $L(\Omega)$ is obtained by estimating Equation 12.3 by maximum likelihood. For large samples, the LR statistic is distributed approximately as a χ^2 with 2 degrees of freedom. The number of degrees of freedom is the number of restrictions imposed by the hypothesis. For this test, the number of degrees of freedom equals the number of parameters constrained to be the same. By the LR test, the hypothesis is rejected at significance level α if LR > $\chi^2(n)$, where $\chi^2(n)$ is the upper significance point of a χ^2 distribution with n degrees of freedom.

The hypothesis of equality given parallelism is H(E|P): $\alpha_s = \alpha_f$. The unrestricted model is Equation 12.3 and the restricted model is

$$y = \alpha + \beta_1 x_1 + \beta_2 x_2. \tag{12.4}$$

In Equation 12.4, the coefficients for spray and film are restricted to be the same. The maximum log likelihood values required for this test are $L(\Omega)$ (the ML estimate of Equation 12.3) and $L(\omega)$ (the ML estimate of Equation 12.4). When H(E|P) is not rejected, $LR = 2[L(\Omega) - L(\omega)] \sim \chi^2(1)$. The hypothesis is rejected at significance level if $LR \leq \chi^2(1)$. Once H(P) is not rejected, H(E|P) can be tested.

The hypothesis of equality is H(E): $\alpha_s = \alpha_f$, $\beta_{1s} = \beta_{1f}$, $\beta_{2s} = \beta_{2f}$. The unrestricted models are Equations 12.1 and 12.2, and the restricted model is Equation 12.4. The maximum log likelihood values required are $L(\Omega)$ and $L(\omega)$, where $L(\Omega) = L_s + L_f$ (i.e., ML estimations of Equations 12.1 and 12.2) and $L(\omega)$ is the maximum likelihood estimation of Equation 12.4. When H(E) is not rejected, $LR = 2[L(\Omega) - L(\omega)] \sim \chi^2(3)$.

12.5.3 Data Analysis with PoloMulti

PoloMulti[4] is the Microsoft Windows- and Apple OS-compatible version of the older program PoloEncore,[3] but PoloMulti is vastly improved in user friendliness and easy comprehension of results. The program is designed to analyze the effects of up to nine explanatory variables on a binary response (e.g., dead vs. alive). Figure 12.1a and b shows the PoloMulti screens that produce the solutions to Equations 12.1 through 12.4. Solutions to the hypothesis tests are shown in Figure 12.2, which is actually the output screen. The hypotheses and results of testing are

(1) H(P) [Hypothesis of parallelism]
      $L(\Omega) = -223.8762$
      $L(\omega) = -225.1922$
      $LR = 2.6321$; df = 2; The hypothesis is rejected at P = 0.05 if $LR > X^2 = 5.99$. Because $2.6321 < 5.99$, the hypothesis cannot be rejected. In this case, H_0 cannot be rejected. Dr. Paula Maven concludes that the planes (i.e. lines) are parallel.

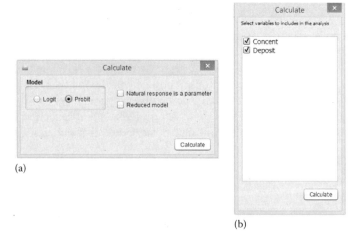

(a)

(b)

Figure 12.1   (a) Parameter selection screen. (b) Variable selection screen.

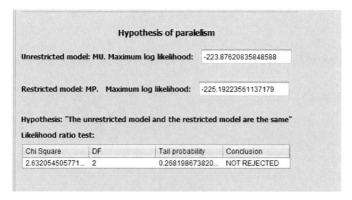

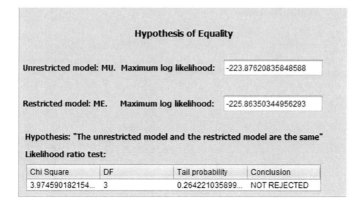

Figure 12.2   PoloMulti output for Tattersfield and Potter study showing LR tests.

(2) H(E|P) [Hypothesis of equality given parallelism]

$L(\Omega) = -225.1922$

$L(\omega) = -225.8635$

LR = 1.3425; df = 1; This hypothesis must be rejected at P = 0.05 if LF > X^2 = 3.84. H_0 cannot be rejected, and Paula concludes that the planes are parallel and equal.

(3) H(E) [Hypothesis of equality]

$L(\Omega) = -223.8762$

$L(\omega) = -225.8635$

LR = 3.9746; df = 3; This hypothesis must be rejected at P = 0.05 if LR > X^2 = 7.81. H_0 cannot be rejected, and Paula concludes that the planes are equal.

12.6 MULTIPLE LOGIT ANALYSIS OF DOSE–WEIGHT–TEMPERATURE–PHOTOPERIOD–RESPONSE DATA WITH R

An unpublished study in which nine groups of experiments were done to study the effects of temperature and photoperiod on response of the western spruce budworm to pesticides demonstrates estimation of multiple logit regressions by addition of categorical covariates to the linear predictor. In this study, three temperatures—10°C, 24°C, and 28°C—and three photoperiods—8:16, 12:12, and 16:8 (light [L]:dark [D])—were tested. The average body weight of each treatment group (10 last instars in a Petri dish lined with filter paper) was determined. Each experiment was replicated three or four times.

12.6.1 Statistical Model

The logit model that expresses the effect of the two categorical variables (temperature and photoperiod) and two continuous variables (dose and weight) is

$$y_{ij} = \alpha_{ij} + \beta_{ij} \log_{10}(d + d_0) + \gamma_{ij} \log_{10}(W), \qquad \text{(Model 1)}$$

where y_{ij} is the logit of the probability of mortality, $i = 1, 2, 3$ corresponds with the three temperature levels and $j = 1, 2, 3$ corresponds with the three photoperiods. The term d is dose in micrograms, W is average body weight in milligrams, and $d_0 = 10^a$, with

$$a = \log_{10}(d_0) = \log_{10}(d_1 - \frac{d_1}{d_2 - d_1}[\log_{10}(d_2) - \log_{10}(d_1)]$$

and $0 < d_1 < ... < d_l$. This method for dealing with dose zero is discussed by Tukey et al.[20] This model contains 27 (3 x 3 x 3) parameters and is equivalent to fitting separate logit planes to each of the nine groups.

Many models could be fit to these data. For example, the logit plane

$$y_{ij} = \beta \log_{10}(d + d_0) + \gamma \log_{10}(W) \qquad \text{(Model 2)}$$

with 11 parameters fits parallel logit planes to the nine groups. In this case, the slopes β and y do not change when temperature, photoperiod, or both change.

The logit line

$$y_{ij} = \alpha_i + \beta_i \log_{10}(d + d_0) \qquad \text{(Model 3)}$$

with six parameters fits a model with no effects of weight or photoperiod. Here, the slopes (β_i) and the intercept (α_i) depend on the three temperature levels but do not depend on the photoperiods ($j = 1, 2, 3$).

12.6.2 Hypothesis Tests

The first step in hypotheses testing consists of fitting the logit model as specified by Model 1 with a computer program such as R. Next, the restricted models with fewer parameters (e.g., Models 2 and 3) are fit. The LR test statistic is then calculated to test various hypotheses about the general Model 1 based on the difference in values of the maximum log likelihoods (or deviances) of various models.

For example, the difference in the deviances of the models given by Models 1 and 2 can be used to test the hypothesis of parallelism of the logit planes. If the LR test statistic is larger than the corresponding value from a X^2 table, the hypothesis is rejected. The df for a particular test is the difference between the number of parameters for the models.

Use of the $P = 0.01$ level of significance is appropriate because more than one hypothesis will usually be tested on the same data set. Testing at this low level will probably ensure that the overall P value will not be $> (1 - 0.99k)$, where k is the number of hypotheses tested. For example, if $k = 5$, then the overall P value of the experiment will not be >0.049 (i.e., about 5%).

12.6.3 Search for the "Best-Fitting" Dose–Mortality Model

The LR test statistic can be used to search for the "best-fitting" logit (or probit) plane, given the set of covariates (e.g., temperature, body weight) recorded in the experiment. The search method consists of computing a sequence of logit planes starting with the model for separate planes given by Model 1, then fitting a model with a smaller subset of parameters at each consecutive step. The search is complete when any further reduction in the parameters produces a significant LR test statistic.

12.6.4 Example: Acephate

One of the pesticides that was tested in the temperature–photoperiod–dose–response experiment was acephate. Some of the data from this experiment are shown in Figure 12.3a. Under the TEMP column, 1, 2, and 3 designate 10°C, 24°C, and 28°C, respectively. In the PHOTO column, 1, 2, and 3 designate photoperiods of 8:16, 12:12, and 16:8 (L:D), respectively. The analyses were done with R. Figure 12.3b shows the commands for testing the various hypotheses; Figure 12.3c lists the information necessary for calculating the LR test statistics. Several hypotheses are of interest, as described in the following.

	dose	response	weight	total	temp	photo
1	0	0	90	10	1	1
2	0	0	70	10	1	1
3	0	0	80	10	1	1
4	0	0	70	10	1	1
5	0	0	90	10	1	2
6	0	0	70	10	1	2
7	0	0	80	10	1	2
8	0	0	100	10	1	2
9	0	0	80	10	1	3
10	0	0	70	10	1	3
11	0	0	90	10	1	3
12	0	0	70	10	1	3
13	0	0	90	10	2	1
14	0	0	80	10	2	1
15	0	0	100	10	2	1
16	0	0	90	10	2	1
17	0	0	110	10	2	2
18	0	0	80	10	2	2
19	0	0	80	10	2	2
20	0	0	100	10	2	2
21	0	0	70	10	2	3
22	0	0	70	10	2	3
23	0	0	90	10	2	3
24	0	0	70	10	2	3
25	0	0	80	10	3	1
26	0	0	90	10	3	1

(a)

Figure 12.3 Multiple logit analysis of dose–weight–temperature–photoperiod–response data with R. (a) Input data for acephate and western spruce budworm.   (*Continued*)

```
1        setwd('C:/users/Braad/Documents/Entomology/data/')
2        DATA=read.csv('DATAch12new.csv',header=T)
3        colnames(DATA)=c('DOSE','RESPONSE', 'WEIGHT','TOTAL', 'TEMP', 'PHOTO')
4
5        attach(DATA,2)
6
7        TEMP=as.factor(TEMP)
8        PHOTO=as.factor (PHOTO)
9        r=RESPONSE/TOTAL
10       log.dose=log10(DOSE+.008)
11       log.weight=log10(WEIGHT)
12
13       all.possible-model.matrix(r~ PHOTO*log.dose*TEMP +
14                          PHOTO*log.weight*TEMP-1,
15                            Contrasts.arg =
16                              Lapply(cbind.data.frame(PHOTO,TEMP),
17                                  Contrasts, contrasts=FALSE))
18
19       #matrix of data for each of the 4 models
20       mod.matl=all.possible[,c(12:30, 30:47)]#separate lines and weight
21       mod.mat2=all.possible[c(12:20, 30:38)]#separate lines no weight
22       mod.mat3=all.possible[,c(4,12:20)]#parallel lines no weight
23       .
24       #model1
25       mod1=glm(r~mod.mat1-1 , binomial(link='logit'),weight=TOTAL)
26       het=mod1$dev/mod1$df.residual
27       print(paste('HETEROGENEITY FACTOR =' , round(het,4)))
28       summary(modd1)
29
30       #model2
31       mod2=glm(r~mod.mat2 -1 ,binomial (link= 'logit'),weight=TOTAL)
32       summary(mod2,dispersion=het,cor=F)
33
34       #model3
35       mod3=glm(r~mod.mat3 -1 ,binomial (link= 'logit'),weight=TOTAL)
36       summary(mod3,dispersion=het,cor=F)#best model
37
38       #separate line lines w/weight vs separate lines w/o weight
39       1-pchisq(mod2$deviance-mod1$deviance,mod2$df.residual-mod1$df.residual)
40
41       #separate lines w/o weight vs parallel lines w/o weight
42       1-pchisq(mod3$deviance-mod2$deviance,mod3$df.residual-mod2$df.residual)
(b)
```

Figure 12.3 (Continued)   Multiple logit analysis of dose–weight–temperature–photoperiod–response data with R. (b) R code for models of parallelism and equality.   (*Continued*)

```
1     [1] "HETEROGENEITY FACTOR = 1.8781'
2
3     #separate lines w/weight
4     Null deviance:     2635.74  on 432          degrees of freedom
5     Residual deviance:   760.63  on 405          degrees of freedom
6
7     #separate lines w/o weight
8     Null deviance:  2635.74  on 432     degrees of freedom
9     Residual deviance:   771.96  on 414   degrees of freedom
10
11    #parallel lines w/o weight (best model)
12    Null deviance:  2635.74  on 432  degrees of freedom
13    Residual deviance:  788.92  on 422  degrees of freedom
14
15    #separate lines w/weight vs separate lines w/o weight p-value
16    0.2534623
17
18    #separate lines w/o weight vs parallel lines w/o weight p-value
19    0.03052379
20
21    #Estimates from best model
22    Coefficients:
23                              Estimate      Std. Error      Z value      Pr(>|z|)
24    .mod.mat3log.dose          1.58747       0.07621         20.831       < 2e-16    ***
25    .mod.mat3PHOTO1:TEMP1  -0.53300       0.16413         -3.247       0.001165   **
26    mod.mat3PHOTO2:TEMP1   -0.32351       0.16391         -1.974       0.48417    *
27    mod.mat3PHOTO3:TEMP1   -0.56115       0.16133         -3.478       0.000505   ***
28    mod.mat3PHOTO1:TEMP2   -0.44376       0.16244         -2.732       0.006300   **
29    mod.mat3PHOTO2:TEMP2   -0.48833       0.15948         -3.062       0.002198   **
30    mod.mat3PHOTO3:TEMP2   -0.48578       0.15869         -3.061       0.002204   **
31    mod.mat3PHOTO1:TEMP3   -0.51100       0.15851         -3.224       0.001265   **
32    mod.mat3PHOTO2:TEMP3   -0.26674       0.16078         -1.659       0.097113   .
33    mod.mat3PHOTO3:TEMP3   -0.26674       0.16078         -1.659       0.097113   .
(c)
```

Figure 12.3 (Continued)   Multiple logit analysis of dose–weight–temperature–photoperiod–response data with R. (c) results of R code.

12.6.4.1 Significance of Average Body Weight

A test of this hypothesis involves comparison of the deviances for Model 1 and the model with the logit line

$$y_{ij} = \alpha_{ij} + \beta_{ij}\log_{10}(d + d_0).$$ (Model 4)

The differences in the deviances between the two models is $771.96 - 760.63 = 11.33$ with 9 df (see Figure 12.3c). Therefore, the weight covariate does not appear to significantly affect response of the larvae to acephate ($P = 0.25$).

Table 12.2   Estimates of LD$_{90}$, Intercepts (α), and Slope (β) for Western Spruce Budworm Held for Different Photoperiods (L:D) and Temperatures (°C) after Treatment with Acephate

Temp.	L:D	$\alpha \pm$ SE	$\beta \pm$ SE	LD$_{90}$
10	8:16	−0.53 ± 0.16	1.6 ± 0.08	7.12
	12:12	−0.32 ± 0.16	1.6 ± 0.08	5.26
	16:8	−0.56 ± 0.16	1.6 ± 0.08	7.43
24	8:16	−0.44 ± 0.16	1.6 ± 0.08	6.26
	12:12	−0.49 ± 0.16	1.6 ± 0.08	6.72
	16:8	−0.49± 0.16	1.6 ± 0.08	6.72
28	8:16	−0.51 ± 0.16	1.6 ± 0.08	6.92
	12:12	−0.27 ± 0.16	1.6 ± 0.08	4.90
	16:8	−0.27 ± 0.16	1.6 ± 0.08	4.90

12.6.4.2 Parallelism of the Logit Lines

The differences in the deviances between the model specified in Equation 12.4 and the model for parallel logit lines given by

$$y_{ij} = \alpha_{ij} + \beta \log_{10}(d + d_0) \qquad \text{(Model 5)}$$

is 788.92 − 771.96 = 16.96 with 8 df (Figure 12.3c). So, the hypothesis of logit lines being parallel cannot be rejected ($P > 0.01$). Thus, there is not enough evidence in the data to suggest that the slope of the logit line is affected by changing temperatures, photoperiods, or both.

12.6.4.3 Model with the Best-Fitting Logit Line

$$y_{ij} = \alpha_{ij} + \beta_i \log_{10}(d + d_0) \qquad \text{(Model 6)}$$

In this model, weight is not included, and the slope does not depend on temperature or photoperiod. Estimates of the parameters in Model 6 are listed in lines 24–33 of Figure 12.3c. Values of LD$_{90}$ (Table 12.2) were calculated by LD$_{90}$ = 10^a, where $a = (x - \hat{\alpha})/\hat{\beta}$ and $x = \log_e(- \log_e(1 - 0.9)) = 0.834$ and $\hat{\alpha}$ and $\hat{\beta}$ are estimates of the parameters.

12.7 CONCLUSIONS

Quantal response bioassays with multiple explanatory variables are far more efficient in demonstrating the effects of variables other than dose on response. Single variable experiments waste time, test subjects, and financial resources. Reliable computer programs—POLO2, PoloJR, and PoloMulti—are available for multiple regression analyses. Estimation can be done with categorical or continuous variables. Use of a user-friendly program such as PoloMulti is preferable for most users with no interest or skills in experimental statistics.

REFERENCES

1. Robertson, J. L. and Preisler, H. K., *Pesticide Bioassays with Arthropods*, CRC Press, Boca Raton, FL, 1992.
2. Finney, D. J., *Probit Analysis*, Cambridge University Press, Cambridge, England, 1971.
3. LeOra Software, *PoloEncore*, LeOra Software, 1007 B St., Petaluma, CA 94952, 2006.
4. LeOra Software LLC, *PoloMulti*, LeOra Software LLC, PO Box 582, Parma, MO 63870 http://www.LeOra-Software.com.
5. Gast, R. T., The relationship of weight of lepidopterous larvae to effectiveness of topically applied insecticides, *J. Econ. Entomol.* 52, 1115, 1959.
6. Gast, R. T., Guthrie, F. E., and Early, J. D., Laboratory studies on *Heliothis zea* (Boddie) and *H. virescens* (F.), *J. Econ. Entomol.* 49, 408, 1956.
7. MacCuaig, R. D., Determination of the resistance of locusts to DNC in relation to their weight, age, and sex, *Ann. Appl. Biol.* 44, 634, 1956.
8. Way, M. J., The effect of body weight on the resistance to insecticides of last-instar larva of *Diataraxia oleracea*, the tomato moth, *Ann. Appl. Biol.* 41, 77, 1954.
9. Domencich, T. A. and McFadden, D., *Urban Travel Demand*, American Elsevier, New York, 1975.
10. Russell, R. M., Savin, N. E., and Robertson, J. L., POLO2: A user's guide to multiple probit or logit analysis, USDA Forest Service Gen. Tech. Rep., PSW-55, 1981.
11. Maddala, G. S., *Econometrics*, McGraw-Hill, New York, 1977.
12. LeOra Software, *PoloEncore*, LeOra Software, 1007 B St., Petaluma, CA 94952, 2006.
13. Reichenbach, N. C. and Collins, W. J., Multiple logit analyses of temperature and humidity on the toxicity of propoxur to German cockroaches (Orthoptera: Blatellidae) and western spruce budworm (Lepidoptera: Tortricidae), *J. Econ. Entomol.* 77, 31, 1984.
14. Brown, M., ed., *Biomedical Computer programs, P Series*, University of California, Los Angeles, 1981.
15. Baker, J. R. and Nelder, J. A., *The GLIM Manual: Release 3. Generalized Linear Interactive Modelling*, Numerical Algorithms Group, Oxford, England, 1978.
16. Shapiro, M., Preisler, H. K., and Robertson, J. L., Enhancement of baculovirus activity on gypsy moth (Lepidoptera: Lyantriidae) by chitinase, *J. Econ. Entomol.* 80, 1113, 1987.
17. Robertson, J. L., Russell, R. M., Preisler, H. K., and N.E. Savin. *Bioassays with Arthropods*, CRC Press, Boca Raton, FL, 2007.
18. Tattersfield, F. and Potter C., Biological methods of determining the insecticidal values of pyrethrum preparations (particularly extracts of heavy oil), *Ann. Appl. Biol.* 30, 259, 1943.
19. Busvine, J. R., *A Critical Review of the Techniques for Testing Insecticides*, Commonwealth Agricultural Bureaux, London, 1971.
20. Tukey, J. W., Ciminera, J. L., and Heyse, J. F., Testing the statistical certainty of a response to increasing doses of a drug, *Biometrics* 41, 295, 1985.

CHAPTER 13

Multiple Explanatory Variables: Body Weight

Jessica has been rearing a laboratory colony of *Patronius giganticus* for 18 months, and Dr. Maven has observed that the insects have become heavier with each successive generation. For the first few weeks of bioassays, Jessica selected and tested insects within a very narrow weight range. However, at month 3, the insects' weights have increased so much that they are nowhere close to what they were in the beginning, and Jessica simply cannot find enough insects in the weight range that she had used previously. A protocol change is in order, so she brings the matter to Paula's attention. Dr. Maven ponders this situation—what should they do? Jessica could conduct another set of bioassays using a weight range with an upper limit, but what if the weight change affects the pesticide efficacy? If it does, all of her previous results will have lost their value for comparative purposes. Once again, an online literature search for similar research reveals that multiple probit or logit regression might be the most efficient method to handle this burdensome problem. Just in case weight is a significant factor, Jessica will record binary responses (dead or alive), doses, and body weights for each larva. This type of bioassay is a binary quantal response experiment with multiple explanatory variables.

The assumption that insects generally respond to toxicants in direct proportion to their body weight has led some investigators[1,2] to compensate for effects of weight by proportionally adjusting doses to the weight of the test subjects during the application process. The adjustment (e.g., application of 1 μl/100 mg body weight) was (and unfortunately, still is) used routinely. Results of experiments with locust species[3] and several species of Lepidoptera[1,4] suggested that such dose adjustments were appropriate. But evidence that responses do not vary as simple functions of body weight was also described by Bliss[5] and Way.[6]

Evidence, summarized first by Busvine[7] and later by Robertson et al.,[8] should be sufficient to alert any investigator about the possible significance of weight in the response of a particular arthropod species and the fact that the significance of the weight variable should be tested before any adjustment is made. Yet many researchers do not do so, either because they do not know how or because weight adjustment has been used so often in the past. General publications after 1979 emphasize methods to deal with body weight as a variable. In publications prior to 1979,[9–33] investigators applied 1 μl/100 mg body weight no matter what the pesticide and regardless of the species being tested with the assumption that responses of insects are proportional to body weight.

Dr. Maven is slightly skeptical about proportional response occurring in many moth pests, including western spruce budworm (Lepidoptera: Tortricidae). Before 1979, for example, response to approximately 300 different pesticides was proportional to body weight of western spruce budworm. After 1979, proportional response abruptly ceased. Was there something in trees that caused proportional response of western spruce budworm to occur only before 1979? A terpene or two? What else could account for this strange phenomenon? Simple: Dr. Maven concludes that this scientist was probably just following contemporary convention. Then she begins to wonder how many others still blindly follow one convention or another without thinking about what they are doing. After examining some recent papers dealing with pesticide toxicity, she suspects that there may yet be quite a few. Anyone who has spent the time rearing arthropods in a laboratory culture or collecting them from a field population cannot afford to compromise the bioassay by not identifying and testing hidden assumptions in either the application method or in the data analysis. In particular, any a priori assumption about the role of body weight in response to pesticides must be avoided.

13.1 EFFECTS OF ERRONEOUS ASSUMPTIONS ABOUT BODY WEIGHT

In Dr. Maven's opinion, two publications should be required reading for any investigator who is doing a topical application bioassay. In the first report, the authors[8] tested the hypothesis that the response of western spruce budworm to mexacarbate, pyrethrins, or dichlorodiphenyltrichloroethane (DDT) is proportional to body weight. For each chemical, different sampling methods, weight ranges, and dose adjustment procedures were compared. Tolerance to mexacarbate and DDT appeared to increase as a function of the square of body weight, while in a few instances, tolerance to pyrethrins increased as a simple proportion of body weight. Overall, however, the square of weight more adequately described the relationship between tolerance and weight.

The same data set[34] was used to describe calculation of lethal doses and confidence intervals for insects of a given body weight. Average and individual body weights were used to examine tolerances with and without assumptions that response is proportional to body weight. For these comparisons, calculations were based on probit and logit regression techniques. One of the conclusions of this investigation was that the unquestioned use of proportional adjustment of response for body weight may result in erroneous inferences about relative toxicities. A crossover effect (see Figure 13.1) occurred.

Lethal doses for the 60-mg weight class (the smallest group of western spruce budworm tested) were higher assuming proportional response than when no assumption was made, and lethal doses for the higher weight classes were generally lower assuming proportional response than when no assumption was made. In general, at

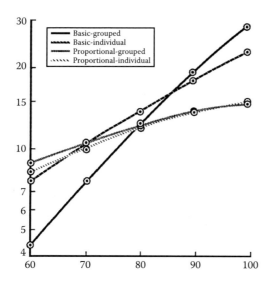

Figure 13.1 Estimated LD_{90}s for DDT applied to last instar western spruce budworm in experiments with four designs. (From Savin NE, Robertson JL, Russell RM, *J. Econ. Entomol.*, 75, 538, 1982.)

lower body weights, use of proportional adjustment for body weight will indicate that a chemical is less toxic, and that at higher weights, it is more toxic, than is actually the case. In addition, confidence limits for the lethal doses were generally larger when no assumption about proportional response was made. In essence, use of the proportional response adjustment appears to bias lethal dose estimates and give an overly optimistic impression of their precision. The results of this investigation with western spruce budworm certainly cannot be generalized to all other arthropods. Increased tolerance with increased weight, for example, was observed in the granary weevil, *Sitophilus granarius* (L.).[35] However, dose adjustment should never be used without testing the hypothesis that response is proportional to body weight. Grouping of test subjects tended to mask the absence of a proportional response in these experiments.[34] Proportional response for weight was rejected in 73% of the experiments in which individual insect data were used but in only 44% of the experiments in which grouped data were used.

In practice, bioassays in which individual test subjects are weighed are rare because they are so time consuming. So, when group weights (such as the average weight of 10 test subjects in a Petri dish) are used to test the hypothesis of proportional response, results must still be viewed with caution even if the hypothesis is not rejected.

Finally, these authors suggest that the most prudent procedure is to use a convenient uniform application volume, such as 1 μl. The hypothesis of proportionality can then be tested by simply including body weight as an additional variable in the multiple probit or logit regression.

13.2 TESTING THE HYPOTHESIS OF PROPORTIONAL RESPONSE

The model with dose and weight as covariates (i.e., proportional response) is

$$y = \delta_0 + \delta_1 \log(D) + \delta_2 \log(W), \qquad (13.1)$$

where y = the probit or logit of the response, D = dose, and W = weight. If $\beta_0 = \delta_0$, $\beta_1 = \delta_1$, and $\beta_2 = \delta_1 + \delta_2$, Equation 13.1 can be rewritten as

$$y = \beta_0 + \beta_1 \log(D/W) + \beta_2 \log W. \qquad (13.2)$$

The hypothesis of proportionality is H: $\beta_2 = 0$. The hypothesis assumes that proportion response is

$$y = \beta_0 + \beta_1 \log(D/W). \qquad (13.3)$$

The hypothesis of proportional response is rejected at the $P = 0.05$ significance level if the likelihood ratio (LR) test statistic

$$LR = 2[L(\Omega) - L(\omega)] > \chi\alpha^2,$$

where $L(\Omega)$ is the maximum log likelihood for Equation 13.2 and $L(\omega)$ is the maximum log likelihood for Equation 13.3. The test statistic LR can be calculated after fitting Equations 13.2 and 13.3 with any of the computer programs (*PoloEncore*[36] or PoloMulti[37]) mentioned in Chapter 12. In *PoloMulti*, calculations for this hypothesis test are when the options in Figure 13.2a–d are chosen.

Figure 13.3 shows the results for Dr. Maven's data for the Godfather larvae analyzed with PoloMulti. Besides the constant term (i.e., intercept), the full model includes the variables of natural response, logarithmic transformations of body weight, and logarithmic transformations of dose divided by weight (dose over weight, or D over W). Parameter values (lines 5–7), their standard errors (lines 9–11), and values of each t-ratio (parameter value ÷ S.E.) (lines 12–14) follow. (If the value of a t-ratio for any term is below the critical value of 1.96, it is not significant in the regression.) The prediction success table (lines 22–27) lists the number of individual test subjects that were predicted to respond and that actually did so, predicted to respond but that actually did not, predicted to be unresponsive and that actually responded, and predicted to be unresponsive and that actually were unresponsive. These groups are listed in increments of 0.00 to 0.50 and 0.50 to 1.00. These numbers are calculated by using maximum probability as a criterion.[38]

The results of the LR test of the hypothesis are given next, in lines 30–32. The value of χ^2 (= 107.37) indicates that the hypothesis that the logarithm of body weight and the logarithm of dose over weight are significant variables in the response. The LR statistic for the test of proportional response is

$$LR = 2[L(\Omega) - L(\omega)] = 2[-94.6799 + 114.0303]$$
$$= 2[19.3504] = 38.7008.$$

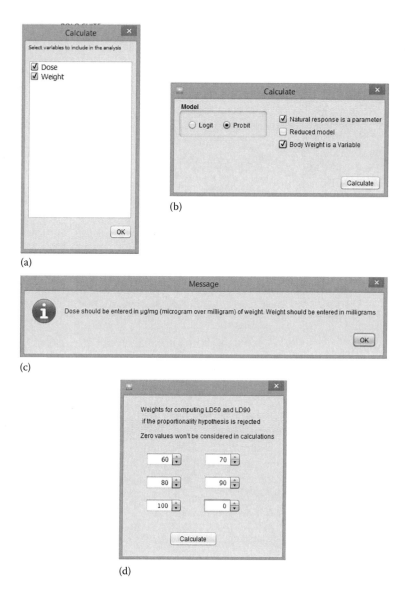

Figure 13.2 (a) PoloMulti options for the proportional response hypothesis testing and for confidence limits and lethal doses (LDs) for specified weights. (b) The proportional response hypothesis testing for confidence limits and lethal doses (LDs) for specified weights. (c) Reminder message that dose and weight should be in microgram/milligram and grams, respectively. (d) Selected weights for computing the LD_{50} and LD_{90} if the proportionality hypothesis test is rejected.

Dr. Maven rejects the hypothesis of proportionality because the critical value for χ^2 at $P = 0.05$ significance level is with 1 df is 3.84. She also can reject the hypothesis on the basis of the t-ratio for β_2 (= −5.561). These results indicate that she cannot justify adjustment for dose by body weight and that she should use the model with weight and dose as independent variables for her analyses. The hypothesis can also be tested with the t-ratio for β_2. If the hypothesis is not rejected, LD_{50} and LD_{90} estimates can be obtained for any body weight desired. If the hypothesis is rejected, two options are available, as described in the next section.

13.3 WHEN BODY WEIGHT IS A SIGNIFICANT INDEPENDENT VARIABLE

If the hypothesis of proportional response has been tested and rejected, another model can be used to estimate selected lethal doses at specific body weights. The linear model for body weight and dose as independent variables[8,34] is

$$y = \delta_0 + \delta_1(\log D) + \delta_2(\log W). \tag{13.4}$$

For $W = W^*$, where W^* is a specified weight, the equation can be rewritten as

$$y = A + B(\log D), \tag{13.5}$$

where $A = \delta_0 + \delta_2 (\log W^*)$ and $B = \delta_1$.

Equation 13.5 can be used to estimate the lethal dose for body weight W^* by simple linear regression. Confidence limits for the lethal dose, given body weight W^*, can then be calculated because A and B are linear functions of the original coefficients δ_0, δ_1, and δ_2. Details of the calculations involved are described by Savin et al.[34] and by LeOra Software LLC.[37]

The multiple regression equation for the data is

$$y = 15.9899 + 3.30682(\log D) - 7.1859(\log W)$$

(see Figure 13.3, line 71). Lines 97–173 in Figure 13.3 list the lethal dose estimates for Godfather larvae of average weight (77.367 mg) and the body weights that Dr. Maven has specified (60, 70, 80, 90, and 100 mg).

13.4 STANDARDIZED BIOASSAY TECHNIQUES INVOLVING WEIGHT

Use of insects in a very narrow weight range has been recommended in standardized tests for resistance (or relative susceptibility) of the *Heliothis* species.[39] This recommendation has some merit because it helps ensure uniformity among different investigators testing different test subjects in different laboratories.

1	UNRESTRICTED PROPORTIONAL MODEL WITH 3 PARAMETERS

2	Parameter		Value	
3	Null Log Likelihood (parameters = zero) -148.3665			
4	Maximum Log Likelihood		-94.6799	

5 Parameters

6	Const	LogWeight	LogDoverW	Natural
7	18.1481	8.3306	6.1828	0.0148
8				
9	Standard Error			
10	Const	LogWeight	LogDoverW	Natural
11	2.982	1.498	0.8024	0.0146

12	t ratio			
13	Const	LogWeight	LogDoverW	Natural
14	6.0861	-5.5611	7.7057	1.0096

15 Covariances

16		Const	LogWeight	LogDoverW	Natural
17	Const	8.8921	-4.4462	1.3032	0.0046
18	LogWeight	-4.4462	2.244	-0.563	-0.0024
19	LogDoverW	1.3032	-0.563	0.6438	8.0e-4
20		0.0046	-0.0024	8.0E-4	2.0E-4
21					

22 Prediction success table

23		Observed	Observed	Observed
24	Probability	Non-responding	Responding	Fraction Responding
25	0.000-0.500	72.00	23.0	0.2421
26	0.500-1.00	24.0	88.0	0.7857

27	Prediction percentage	77.2947

28 Hypothesis tested at 5% significance level: "The parameter model and the 0-parameter
29 model are the same"

30 Likelihood Ratio Tests

31	Chi Square	DF	Tail Probability	Conclusion
32	107.3731	4.0	0.0	REJECTED

33	RESTRICTED PROPERTIONAL MODEL WITH 2 PARAMETERS

34	Parameter	Value
35	Null Log Likelihood (parameters=zero) -148.3393	
36	Maximum Log Likelihood	-114.0303

37 Parameters

38	Const	LogDoverW	Natural
39	1.8539	4.677	0.0162

(a)

Figure 13.3 PoloMulti results of calculations produced by options chosen in Figure 13.2a–d.
(*Continued*)

40	Standard Error		

41	Const	LogDoverW	Natural
42	0.2561	0.6438	0.016

43	t ratio		

44	Const	LogDoverW	Natural
45	7.2398	7.2646	1.0122

46 Covariance

47		Const	LogDoverW	Natural
48	Const	0.0656	0.1513	0.0
49	LogDoverW	0.1513	0.4145	8.0E-4
50	Natural	-0.0	8.0E-4	3.0E-4

51 Prediction Success Table

52		Observed	Observed	Observed
53	Probability	Non-responding	Responding	Fraction Responding
54	0.000-0.500	58.0	12.0	0.1714
55	0.500-1.00	38.0	99.0	0.7226

56	Predicted percentage	75.8455

57 Hypothesis tested at 5% significance level: "The 2 parameter model and the
58 0-parameter model are the same"

59 Likelihood Ratio Tests

60	Chi Square	DF	Tail Probability	Conclusion
61	68.618	3.0	0.0	REJECTED

62 Hypothesis of Proportionality

63 Hypothesis tested at 5% significance level: "The 3-parameter model and the
64 2-parameter model are the same"

65 BASIC MODEL WITH 3 PARAMETERS

66	Parameter	Value
67	Null Log Likelihood (parameters=zero)	-148.3642
68	Maximum Log Likelihood	-94.9087

69 Parameters

70	Const	LogDose	LogWeight	Natural
71	15.9899	3.3068	-7.1859	0.0149
72				

73	Standard Error			
74	Const	LogDose	LogWeight	Natural
75	2.8438	0.4307	1.4362	0.0147

(b)

Figure 13.3 (Continued)   PoloMulti results of calculations produced by options chosen in Figure 13.2a–d.
(*Continued*)

76	t ratio			
77	Const	LogDose	LogWeight	Natural
78	5.6227	7.6783	-5.0034	1.0096

79	Covariances			
80		Const	LogDose	LogWeight	Natural
81	Const	8.0872	0.5907	4.0619	0.0046
82	LogDose	0.5907	0.1855	-0.2445	5.0E-4
83	LogWeight 4.0619	0.2445	2.0627	0.0025
84	Natural	0.0046	5.0E-4	-0.0025	2.0E-4

85   Prediction Success Table

86		Observed	Observed	Observed
87	Probability	Non-responding	Responding	Fraction Responding
88	0.000-0.500	72.0	23.0	0.2421
89	0.500-1.00	24.0	68.0	0.7857

90   Predicted percentage   77.2947

91   Hypothesis tested at 5% significance level: "The 3-parameter model and the
92   0-parameter model are the same"

93   Likelihood Ratio Tests

94	Chi Square	DF	Tail Probability	Conclusion
95	106.9111	4.0	0.0	REJECTED

96   Weight Study Output

97   Weight: 77.3671   log base 10: 1.8886

98   Beta

99	Const	2.41886
100	LogDoverW	3.30682

101   Standard Error

102	Const	2.8438
103	LogDoverW	0.43067

104

105   Covariance

106		Const	LogDoverW
107	Const	0.10202	0.12888
108	LogDoverW	0.12888	0.18548

109   Weight: 60   log base 10: 1.7782

110   Beta

111	Const	3.21223
112	LogDoverW	3.30682

(c)

Figure 13.3 (Continued)   PoloMulti results of calculations produced by options chosen in Figure 13.2a–d.   *(Continued)*

113   Standard Error

114   Const   2.8438
115   LogDoverW   0.43067

116   Covariance

117   Const   LogDoverW
118   Const   0.16389 0.15588
119   LogDoverW   0.15588 0.18548

120   LD50: 0.10681   Lower limit: 0.0799   Upper Limit: 0.1369
121   LD90: 0.26071   Lower limit: 0.2007   Upper Limit: 0.3677

122   Weight: 70   log base 10: 1.8451

123   Beta

124   Const   2.73115
125   LogDoverW   3.30682

126   Standard Error

127   Const   2.8438
128   LogDoverW   0.43067

129   Covariance

130   Const   LogDoverW
131   Const   0.12037 0.13951
132   LogDoverW   0.13951 0.18548

133   LD50: 0.14931   Lower limit: 0.1231   Upper Limit: 0.1769
134   LD90: 0.36445   Lower limit: 0.2944   Upper Limit: 0.4991

135   Weight: 80   log base 10: 1.9031

136   Beta

137   Const   2.31443
138   LogDoverW   3.30682

139   Standard Error

140   Const   2.8438
141   LogDoverW   0.43067

142   Covariance

143   Const   LogDoverW
144   Const   0.09762 0.12533
145   LogDoverW   0.12533 0.18548

146   LD50: 0.19957   Lower limit: 0.1692   Upper Limit: 0.2336
147   LD90: 0.48714   Lower limit: 0.3906   Upper Limit: 0.6827

148   Weight: 90   log base 10: 1.9542

149   Beta

(d)

Figure 13.3 (Continued)   PoloMulti results of calculations produced by options chosen in
Figure 13.2a–d.   (*Continued*)

| 150 | Const | 1.94685 |
| 151 | LogDoverW | 3.30682 |

| 152 | Standard Error |

| 153 | Const | 2.8438 |
| 154 | LogDoverW | 0.43067 |

| 155 | Covariance |

156		Const	LogDoverW
157	Const	0.08907	0.11282
158	LogDoverW	0.11282	0.18548

| 159 | LD50: 0.25779 | Lower limit: 0.2113 | Upper Limit: 0.3165 |
| 160 | LD90: 0.62923 | Lower limit: 0.4849 | Upper Limit: 0.9305 |

| 161 | Weight: 100 | log base 10: 2.0 |

| 162 | Beta |

| 163 | Const | 1.61804 |
| 164 | LogDoverW | 3.30682 |

| 165 | Standard Error |

| 166 | Const | 2.8438 |
| 167 | LogDoverW | 0.43067 |

| 168 | Covariance |

169		Const	LogDoverW
170	Const	0.09056	0.10164
171	LogDoverW	0.10164	0.18548

| 172 | LD50: 0.32411 | Lower limit: 0.2507 | Upper Limit: 0.4269 |
| 173 | LD90: 0.79112 | Lower limit: 0.5783 | Upper Limit: 1.2489 |

(e)

Figure 13.3 (Continued) PoloMulti results of calculations produced by options chosen in Figure 13.2a–d.

However, the recommendation that results should be reported in units proportional to body weight is fallacious. For *Heliothis* species, LD_{50}s are to be presented in standard units of 1 µg/g of larval weight. Yet the hypothesis of proportionality has never been tested. Such a recommendation simply helps perpetuate the erroneous assumption that response is proportional to body weight.

Depending on the purpose of the study, body weight can be included in an experiment in several ways, but *never when the pesticide is applied*. Instead, a uniform application volume such as 1 µl per insect should be used. First, a narrow weight range might be used, as recommended for *Heliothis* species. Because weight does not differ among test groups, it can be ignored in estimating dose–mortality regressions.

In studies of resistance, however, use of a narrow weight range may be impossible despite highly controlled conditions. If necessary, test subjects must be weighed individually if valid and meaningful comparisons between or among responses are to be made. For example, Robertson et al.[40] compared responses of light brown apple moth, *Epiphyas postvittana* (Walker), strains that were resistant and susceptible to azinphosmethyl. Both strains were reared on apple, gorse, broom, blackberry, or general purpose diet. Larvae on natural plant foliage, especially on blackberry, were smaller (18 mg) than larvae on general purpose diet (30 mg). The results of one bioassay showed that responses of resistant larvae from blackberry were dissimilar to those susceptible larvae from general purpose diet. Did the difference in response result from the difference in body weight, or was its effect related to the host plant?

Individual larvae were weighed before application of azinphosmethyl. Multiple regressions with dose and body weight were then estimated. Weight was not a significant variable in either regression, indicating that differences in weight were not the cause of the differences in response that were observed. Probit regressions with dose as the single variable were used. In summary, the procedure used was to test the significance of body weight as a variable. Because it was shown not to be significant, body weight was excluded from subsequent analyses.

If a bioassay is done for predictive purposes, use of a narrow weight average cannot be recommended. Ideally, each insect should be weighed. Regressions should be estimated with Equation 13.4, and lethal doses for a range of weights should be estimated with Equation 13.5. If individual weights cannot be recorded, the next preferable alternative is use of average weights for treatment groups. The least preferable alternative is not to weigh any test subjects. In this case, results are appropriate only if the distribution of body weights happens to be similar in all test groups compared.

13.5 CONCLUSIONS

To date, the most important application of multiple regression techniques (multiple probit or logit analysis) to pesticide bioassays has been examination of the role of body weight in response.[8,34] Results of these investigations showed that the unquestioned use of proportional adjustment of dose to body weight may result in erroneous inferences about relative toxicities. Regardless of the arthropod species tested, the hypothesis that body weight *must* be tested before response is expressed in terms of body weight. *Adjustments for weight should never be made when a pesticide is applied.*

Finally, standardized protocols for tests of pesticide resistance must be changed so that response is not expressed in terms of body weight. Instead, response should be expressed on a per insect basis. Either test subjects should be selected within a very narrow weight range, or they should be weighed individually or in groups of uniform size.

REFERENCES

1. Gast, R. T., Guthrie, F. E., and Early, J. D., Laboratory studies on *Heliothis zea* (Boddie) and *H. virescens* (F.), *J. Econ. Entomol.* 49, 408, 1956.
2. Metcalf, R. L., Methods of topical application and injection, in Shepard, H. H., ed., *Methods of Testing Chemicals on Insects*, Burgess Publishing Co., Minneapolis, MN, 1958, p. 92.

3. MacCuaig, R. D., Determination of the resistance of locusts to DNC in relation to their weight, age, and sex, *Ann. Appl. Biol.* 44, 634, 1956.

4. Gast, R. T., The relationship of weight of lepidopterous larvae to effectiveness of topically applied insecticides, *J. Econ. Entomol.* 52, 1115, 1959.

5. Bliss, C. I., The size factor in the action of arsenic upon silkworm larvae, *J. Exp. Biol.* 13, 95, 1936.

6. Way, M. J., The effect of body weight on the resistance of insecticides of last-instar larva of *Diataraxia oleracea*, the tomato moth, *Ann. Appl. Biol.* 41, 77, 1954.

7. Busvine, J. R., *A Critical Review of the Techniques for Testing Insecticides*, Commonwealth Institute of Entomology, London, 1971.

8. Robertson, J. L., Savin, N. E., and Russell, R. M., Weight as a variable in the response of western spruce budworm to insecticides, *J. Econ. Entomol.* 74, 643, 1981.

9. Schwartz, J. L. and Lyon, R. L., Laboratory culture of orange tortrix and its susceptibility to four insecticides, *J. Econ. Entomol.* 63, 1788, 1970.

10. Schwartz, J. L. and Lyon, R. L., Contact toxicity of five insecticides to California oakworm reared on artificial diet, *J. Econ. Entomol.* 64, 146, 1971.

11. Lyon, R. L., Brown, S. J., and Robertson, J. L., Contact toxicity of sixteen insecticides applied to forest caterpillars reared on artificial diet, *J. Econ. Entomol.* 65, 928, 1972.

12. Robertson, J. L., Lyon, R. L., Shon, F. L., and Gillette, N. L., Contact toxicity of 20 insecticides to *Symmerista canicosta*, *J. Econ. Entomol.* 65, 1560, 1972.

13. Robertson, J. L. and Lyon, R. L., Western spruce budworm: Nonresistance to Zectran, *J. Econ. Entomol.* 66, 801, 1973.

14. Robertson, J. L. and Lyon, R. L., Elm spanworm: Contact toxicity of 10 insecticides to larvae, *J. Econ. Entomol.* 66, 627, 1973.

15. Robertson, J. L. and Gillette, N. L., Western tent caterpillar: Contact toxicity of 10 insecticides applied to the larvae, *J. Econ. Entomol.* 66, 629, 1973.

16. Robertson, J. L. and Lyon, R. L., Douglas-fir tussock moth: Contact toxicity of 20 insecticides applied to the larvae, *J. Econ. Entomol.* 66, 1255, 1973.

17. Robertson, J. L., Page, M., and Gillette, N. L., *Calocalpe undulata*: Contact toxicity of 10 insecticides to the larvae, *J. Econ. Entomol.* 67, 706, 1974.

18. Robertson, J. L., Lyon, R. L., and Gillette, N. L., Contact toxicity of 38 insecticides to pales weevil adults, *J. Econ. Entomol.* 68, 124, 1975.

19. Page, M. and Robertson, J. L., Western spruce budworm, bioactivity of mexacarbate during storage, *Insecticide Acaricide Tests* 1, 104, 1975.

20. Savin, N. E., Robertson, J. L., and Russell, R. M., A critical evaluation of bioassay in insecticide research: Likelihood ratio tests of dose-mortality regression, *Bull. Entomol. Soc. Am.* 23, 257, 1977.

21. Robertson, J. L., Boelter, L. M., and Gillette, N. L., Laboratory tests of insecticides on western spruce budworm, 1974–76, *Insecticide Acaricide Tests* 2, 110, 1977.

22. Russell, R. M., Robertson, J. L., and Savin, N. E., POLO: A new computer program for probit analysis, *Bull. Entomol. Soc. Am.* 23, 209, 1977.

23. Robertson, J. L., Gillette, N. L., Lucas, B. A., Savin, N. E., and Russell, R. M., Comparative toxicity of insecticides to *Choristoneura* species, *Can. Entomol.* 110, 399, 1978.

24. Robertson, J. L., Boelter, L. M., Russell, R. M., and Savin, N. E., Variation in response to insecticides by Douglas-fir tussock moth, *Orgyia pseudotsugata* (Lepidoptera: Lymantriidae), populations, *Can. Entomol.* 110, 325, 1978.

25. Robertson, J. L., Laboratory tests of insecticides on western spruce budworm, *Insecticide Acaricide Tests* 3, 148, 1978.

26. Robertson, J. L. and Gillette, N. L., Contact toxicity of insecticides to western pine beetle, *Insecticide Acaricide Tests* 3, 147, 1978.
27. Robertson, J. L., Crisp, C. E., and Look, M., Toxicity of O,0-dimethyl 2,2,2-trichloro-1-hydroxyethyl phosphonate and its acyloxy and vinyl derivatives, *Insecticide Acaricide Tests* 3, 144, 1978.
28. Robertson, J. L. and Boelter, L. M., Toxicity of insecticides to Douglas-fir tussock moth *Orgyia pseudotsugata* (Lepidoptera: Lymantriidae): I. Contact and feeding toxicity, *Can. Entomol.* 111, 1145, 1979.
29. Stock, M. W. and Robertson, J. L., Differential response of Douglas-fir tussock moth, *Orgyia pseudotsugata* (Lepidoptera: Lymantriidae), populations and sibling groups to acephate and carbaryl: Toxicological and genetic analyses, *Can. Entomol.* 111, 1231, 1979.
30. Robertson, J. L. and Kimball, R. A., Toxicities of topically applied insecticides to western spruce budworm, *Insecticide Acaricide Tests* 4, 174, 1979.
31. Robertson, J. L. and Kimball, R. A., Effects of insect growth regulators on western spruce budworm (*Choristoneura occidentalis*) (Lepidoptera: Tortricidae): I. Lethal effects of the last instar treatments, *Can. Entomol.* 111, 1361, 1979.
32. Robertson, J. L. and Kimball, R. A., Toxicities of topically applied insecticides to western spruce budworm, 1979, *Insecticide Acaricide Tests* 5, 203, 1980.
33. Robertson, J. L. and Kimball, R. A., Toxicities of insecticides to Douglas-fir tussock moth, 1979, *Insecticide Acaricide Tests* 5, 202, 1980.
34. Savin, N. E., Robertson, J. L., and Russell, R. M., The effect of body weight on lethal dose estimates for the western spruce budworm, *J. Econ. Entomol.* 75, 538, 1982.
35. Lavadinho, A. M. P., Toxicological studies on adult *Sitophilus granarius* (L.). Influence of individual body weight on the susceptibility to DDT and malathion, *J. Stored Prod. Res.* 12, 215, 1976.
36. LeOra Software, *PoloEncore* and *PoloEncore User's Guide*, 1007 B St., Petaluma, CA 94952, 2007. See http://www.LeOraSoftware.com.
37. LeOra Software, LLC. *PoloMulti*, PO Box 562, Parma, MO 63870 See http://www.LeOra-Software.com.
38. Domencich, T. A. and McFadden, D., *Urban Travel Demand*, American Elsevier, New York, 1975.
39. Anonymous, Standard method for detection of insecticide resistance in *Heliothis zea* (Boddie) and *H. virescens* (F.), *Bull. Entomol. Soc. Am.* 16, 147, 1970.
40. Robertson, J. L., Armstrong, K. F., Suckling, D. M., and Preisler, H. K., Effects of host plants on toxicity of azinphosmethyl to susceptible and resistant light brown apple moth (Lepidoptera: Tortricidae), *J. Econ. Entomol.* 83, 2124, 1990.

CHAPTER 14

Polytomous (Multinomial) Quantal Response

Once Jessica has learned the basics of bioassays, including those with mixtures and multiple variables, her boss, Dr. Paula Maven, decides to write five additional grants—three of which are awarded. Paula assures Jessica that although all the new grants will involve bioassays, she is confident that Jessica will quickly study the new protocols. Dr. Maven has included a study that involves a far more complicated experimental design and statistical analysis than the binary or multiple quantal response analyses that they have done before. This time Jessica treats the Godfather larvae with each of several juvenile hormone analogues (JHAs). Males and females are treated as separate groups. Not only does the chemical cause some larvae to die, but it also causes some to develop into abnormal pupae that die, some into pupae that form adults with major morphological abnormalities, and some into normal pupae which in turn form adults with minor abnormalities.

In this experiment, the explanatory variables are concentrations of the JHA and sex, but multiple response variables (i.e., the various effects caused by the hormone) are also present. Dr. Maven carefully records the following effects as possible results from treatment of Godfather larvae with each JHA:

1. Died as larva before metamorphosis, or in transition between larval and pupal stages.
2. Pupa malformed, died before or during larval-pupal ecdysis.
3. Pupa malformed, adult with major morphological abnormalities.
4. Pupa malformed, adult normal or with only minor morphological abnormalities.
5. Pupa normal, but died before or during pupal-adult ecdysis.
6. Pupa normal, but adult had major morphological abnormalities.
7. Pupa normal, adult normal or with only minor abnormalities.

Just as polytomous logit analysis presents a problem of homonymy for the hearing impaired, polytomous (or multinomial) quantal response bioassays present a problem for most biologists who are not familiar with the statistical methods involved in data analyses. Many investigators collect information about the multiple effects that occur after treatment with classes of pesticides such as JHAs or benzoyl phenylureas, but then do not know how to analyze their data. Either the data remain unanalyzed or the information is collapsed into a binomial structure (dead versus alive) to facilitate use of regular binary probit or logit analysis.

Few examples of analyses of polytomous data are available in the biological literature. Finney[1] described statistical methods for analysis of multiple classification data from research by Gurland et al.[2] In the experiment, insects exposed to guthion residues were classified as dead, moribund, or alive. The situation with three possible outcomes represents the simplest case for which the polytomous model is appropriate. More recently, Hughes et al.[3] described multinomial logit analyses of a large data set in which effects of three types of chemicals on western spruce bud-worm were compared.

In desperation, she once again calls Professor Garland Tarleton for help. Dr. Tarleton suggests a procedure that Dr. Maven can use to analyze her data. Because a program that is user friendly to the biologist has yet to be developed, she will have to use an experimental statistical program for the analysis.

14.1 THE MULTINOMIAL LOGIT MODEL

The example data set for Dr. Maven's experiments is shown in Table 14.1. The Godfather larvae were topically treated with increasing concentrations of the JHA hydroprene in acetone; males and females were tested individually. Each insect responded in one of K categories (i.e., one of the seven effects that Jessica recorded). The statistical analysis will first test whether any of the covariates significantly affect the response. For Dr. Maven's study, the covariates are sex and JHA concentration. Next, the analysis will calculate estimates of the conditional probabilities of response for each of the seven categories.

TABLE 14.1   Sample Data Set for Male or Female Godfather Larvae Exposed to the Insect Growth Regulator Hydroprene, Showing the Number of Responses in Each Category, by Sex and Concentration

Sex	Concentration	Response Category							Total
		1	2	3	4	5	6	7	
Male	0	3	0	0	1	5	6	87	102
	0.1	8	1	2	0	14	10	63	98
	1.0	10	5	3	13	19	3	32	85
	10.0	26	12	3	6	52	5	6	110
Female	0	4	1	1	0	6	19	69	100
	0.1	8	2	0	0	13	21	44	88
	1.0	7	6	2	5	17	14	38	89
	10.0	19	12	8	12	31	11	20	113

Note: Response categories are (1) dead larva; (2) malformed, dead pupa; (3) malformed pupa that developed into an adult with major abnormalities; (4) malformed pupa that developed into a normal adult; (5) normal pupa that died before adult emergence; (6) normal pupa that developed into an adult with major abnormalities; and (7) normal pupa that developed into a normal adult.

Statistical conditional probability depends on a previous occurrence. For example, *Patronius giganticus* molts into a normal pupa; what is the probability that it will then emerge as a normal adult? The condition on which the probability of interest depends is formation of a normal pupa.

14.1.1 Statistical Model

In Dr. Maven's experiment with two explanatory variables, consisting of one continuous variable (dose concentration) and one factor variable (sex), the probability that an insect responds in state k (conditional on its failure to respond in any previous states) is modeled by

$$q_{ijk} = \frac{e^{y_{ijk}}}{1 + \sum_{k=1}^{K-1} e^{y_{ijk}}}, \qquad (14.1)$$

where

$$y_{ijk} = \alpha_{jk} + \beta_{jk} \log(d_i + d_0) \qquad (14.2)$$

and in which q_{ijk} is the probability of response in state k at dose concentration d_i and sex j (where $j = 1, 2$ for male and female respectively) and where α_{jk} is the intercept (one for each state and sex combination), β_{jk} is the slope (one for each state and sex combination), and $d_0 = 10^a$, with

$$a = \log(d_0) = \log(d_i) - \frac{d_1}{d_2 - d_1} [\log(d_2) - \log(d_i)] \qquad (14.3)$$

and $0 < d_1 < d_2 < ... < d_i$. The use of Equation 14.3 as a method for dealing with dose zero is discussed by Tukey et al.[4] The term d_0 is defined as described in Chapter 11 (Section 11.4.2). The number of responses in each dose concentration/sex/state combination, r_{ijk}, out of the total observations in each respective dose concentration/sex combination, nij, are used to estimate the parameters in Equation 14.2. The vector $(r_{ij1}, ..., r_{ijk})$ has a multinomial distribution.

Conditional probabilities can be used to test hypotheses about the effects of explanatory variables and to estimate the unconditional probabilities of response at a particular stage of development. If responses are put in logical and chronological order (i.e., effects on larvae first, effects on pupae second, and so on), then some of the conditional probabilities are also of interest. For example, the conditional probability in response state 5 is the probability that an insect emerging as an adult from a pupa without any deformities will have major morphological abnormalities.

14.1.2 Estimation of Parameters

For estimation of Equation 14.1, r_{ijk} equals the number of insects in category ij (i.e., of sex j that have been treated with dose d_i) that responded with effect k. The variable n_{ijk} equals the number of insects in category ij that were not observed responding in the previous states $1, \ldots, k-1$. In other words, this variable equals the total number of insects in category ij minus all of the insects that responded with effects $1, \ldots, k-1$. The number of responses r_{ijk} out of n_{ijk} are independent binomial variables. Therefore, a software package that performs binomial regressions, such as PoloSuite, can be used to estimate the parameters in Equation 14.1 and compute maximum values of likelihood functions. The significance of the covariates on the probit or logit line can then be tested with likelihood ratio tests. Estimation of the parameters in a multinomial model can be done using the software package R.[5] The program (multinom) that will run multinomial analyses in open source, R, must first be downloaded from the nnet (Neural Network) library (e.g., https://cran.rproject.org/web/packages/nnet /index.html). Multinom can be used to estimate the parameters in Equation 14.1 and compute maximum values of likelihood functions. The significance of the covariates on the probit or logit line such as sex can then be tested with likelihood ratio tests.

14.1.3 Estimation of Response Probabilities

For estimation of response probabilities, the model with the least number of parameters that fits the data adequately is used to calculate estimates of the conditional probabilities of response (P_{ijk}) with effects $k = 1, \ldots, K-1$. For example, if the effect of sex on the responses in all states is found not to be significant, then the model with equal lines for both sexes should be used. This model is $\hat{y}_{ijk} = \alpha_k + \beta_k \log(d_i + d_0)$ and $P_{ijk} = F(\hat{y}_{ijk})$, where

$$F(y) = e^y 1 + e^y \text{ if the logit model is used,}$$
$$= 1 - e^{ey} \text{ if the CLL model is used, or}$$
$$= \Phi(y) \text{ if the probit model is used.}$$

Estimates of the unconditional probabilities, $\hat{q}_{ijk}$, of response with effects $k = 1, \ldots,$ $K-1$ can be calculated using the estimates $\hat{p}_{ijk}$ as follows

$$\hat{q}_{ij1} = \hat{p}_{ij1}, \ \hat{q}_{ij2} = (1 - \hat{p}_{ij1}) * \hat{p}_{ij2},$$

$$\hat{q}_{ij3} = (1 - \hat{p}_{ij1}) * (1 - \hat{p}_{ij2}) * \hat{p}_{ij3}, \text{ and so on,}$$

where conditional probability is $\hat{p}$, and unconditional probability is $\hat{q}$.

The conditional response probability for the last effect K, $\hat{p}_{ijk}$, is equal to 1 and the corresponding unconditional probability, $\hat{q}_{ijk}$, is equal to 1 minus the sum of the

q for all the previous effects. Therefore, the conditional probability for the last effect need not be calculated specifically.

14.1.4 Data Analysis

Analysis of the data (Figure 14.1) with R is shown in Figures 14.2 and 14.3. Column 1 is sex (1,2), dose is column 2, and responses are columns 3–9. Figure 14.2 shows the R commands for reading the data, testing the hypotheses of equality, and producing estimates of the probabilities of response. Figure 14.2, Line 12, 0.077, is the solution to Equation 14.3. Figure 14.3 is the output from Figure 14.2. At the end of line 3, the value 4.34e-06 is the P value for testing the hypothesis of equality of male and female lines. The P value is less than the critical value for χ^2 at the 5% level, so the hypothesis of equality that the logit lines for male and female Godfather larvae are equal is rejected.

Because the responses of males and females are significantly different, parameters of the conditional binomial model with sex as a variable are computed next (see Figure 14.2). Estimates of the parameters with their standard errors are listed in Figure 14.3, lines 7–21. Lines 7–12 show the estimate for the intercepts $\alpha_j k$ (SEX1

```
                        R datach14.txt
1       #sex dose resp
2       1 0    3  0  0 1   5  6 87
3       1 0.1 8  1  2 0  14 10 63
4       1 1     10 5  3 13 19  3 32
5       1 10   26 12 3 6  52  5  6
6       2 0     4  1  1 0   6 19 69
7       2 0.1 8  2  0 0  13 21 44
8       2 1     7  6  2 5  17 14 38
9       2 10   19 12 8 12 31 11 20
```

Figure 14.1 Data from Table 14.1 prepared for the R analysis.

```
1    install.packages('nnet')# need to run this first if not already installed
2    library(nnet)
3
4    #Set working Directory to where data is located
5    setwd('C:/Users/Braad/Documents/Grad school stuff/Entomology/data/')
6    DATA=read.csv('Datach14new.csv',header=F)
7    colnames(DATA)=c('SEX','DOSE',paste('RESP',1:7,sep=''))#name variables
8
9    attach(DATA,2)#make it so that each variable is an object in R session
10
11   RESPONSE=as.matrix(DATA[,3:9])
12   log.dose=log(DOSE + .077)
13   SEX=as.factor(SEX)
14
15   model.sex=multinom(RESPONSE~SEX + SEX:log.dose-1)
16
17   model.null=multinom(RESPONSE~log.dose)
18   summary(model.sex)
19   anova(model.sex,model.null)
20   summary(model.sex,cor=F)
21   print('Fitted values',quote=F)
22   summary(model.sex)$fitted.values
```

Figure 14.2 The R commands for reading the data, testing hypotheses, and estimating probabilities of response.

```
 1  Model                        Resid. df Resid. Dev   Test    Df    LR stat.      Pr(Chi)
 2  1 log.dose                          36   2156.315
 3  2 SEX + SEX:log.dose                 24   2109.139   1 vs 2   12   47.17594   4.345712e-06
 4
 5  Coefficients:
 6                SEX1          SEX2         SEX1:log.dose   SEX2:log.dose
 7  RESP2 -1.2597361  -0.7251200      0.25609457      0.164495933
 8  RESP3 -1.6924203  -1.5810369     -0.12019376      0.329660167
 9  RESP4 -0.8014304  -1.3067190     -0.07294036      0.432783216
10  RESP5  0.6177895   0.5689661      0.03592632     -0.003537959
11  RESP6 -0.6864786   0.4542121     -0.53541236     -0.413173220
12  RESP7  0.8497721   1.3115093     -0.93023236     -0.517736255
13
14  Std. Errors:
15                SEX1       SEX2       SEX1:log.dose   SEX2:log.dose
16  RESP2 0.3781354 0.3079529      0.1891335       0.1565055
17  RESP3 0.3994012 0.4610608      0.2136517       0.2267466
18  RESP4 0.2868728 0.4297940      0.1520472       0.2084160
19  RESP5 0.2023037 0.2111885      0.1058209       0.1093693
20  RESP6 0.2818739 0.2192091      0.1515107       0.1140803
21  RESP7 0.2085522 0.1940916      0.1152750       0.1017673
22
23  Residual Deviance: 2109.139
24  AIC: 2157.139
25
26  Fitted Values:
27        RESP1         RESP2         RESP3        RESP4        RESP5        RESP6        RESP7
28  1 0.03223842 0.004743661 0.008076103 0.017439608 0.05453459 0.06403635 0.81893127
29  2 0.06008051 0.010940753 0.013618013 0.030586509 0.10471735 0.07642654 0.70363033
30  3 0.15517004 0.044870580 0.028309159 0.069247044 0.28858107 0.07506290 0.33875921
31  4 0.22394390 0.114812845 0.031227435 0.084898319 0.45132337 0.03271971 0.06107442
32  5 0.04583097 0.014557126 0.004049895 0.004090252 0.08169536 0.20820596 0.64157050
33  6 0.06367651 0.023193072 0.007403385 0.008147302 0.11317192 0.20509654 0.57931127
34  7 0.11304674 0.055416844 0.023836511 0.031600912 0.19963786 0.17266694 0.40379419
35  8 0.15710184 0.111252444 0.069231542 0.115586026 0.27525192 0.09525624 0.17631999
```

Figure 14.3 The output from running the commands shown in Figure 14.1.

and SEX2) and slopes $\beta_{ij}k$ (SEX1:log.dose and SEX2:log.dose) for states 2–7. The corresponding standard error values are shown in lines 16–21. No values are given for state 1 because the multinomial probabilities are such that $q_{ij1} = 1 - (q_{ij2} + q_{ij3} + \ldots + q_{ijk})$. This means that probabilities for state 1 are obtained by summing the probabilities for all other states and subtracting the result from 1. The sum of the multinomial probabilities over all states equals 1, a consequence of the fact that an insect has to be in one of the seven states. The best model, which contains separate slopes and intercepts for males and females, produces parameter estimates that are used to calculate the probabilities seen from lines 28–35. According to this analysis, the probability of a normal pupa that died before adult emergence (state 5) does not seem to be related to the concentration of hydroprene applied, as the slope estimates for state 5 are very close to 0 (0.0359, −0.0035 for male and female, respectively). Dr. Maven concludes that some other factor besides hydroprene must be responsible for the other malformations, but she will have to design other experiments to identify the causes.

Dr. Maven is also interested in other probabilities, among them whether the probability of a pupa being malformed, i.e., probability of state 2, 3, or 4, increases with increased concentration of hydroprene. Such probabilities can be calculated from the conditional probabilities (Figure 14.3, lines 28–35); by adding the values for each respective dose in columns 2, 3, and 4; and dividing by 2 (for each sex). Dr. Maven finds that the probability of a malformed pupa increases with concentration (for concentration 0.1, $P = 0.047$; for concentration 1.0, $P = 0.126$; for concentration 10, $P = 0.2615$).

14.2 CONCLUSIONS

Use of the polytomous (multinomial) model makes it possible to quantify effects that are masked by collapsing all effects into a binary model in which dead or alive are the only categories (e.g., Robertson and Kimball[6]); it also permits estimation of individual dose–response probit lines for each effect. Polytomous quantal response analyses are more informative, especially for data that concern the effects of JHAs or other chemicals that are not classic poisons. As discussed by Hughes et al.,[3] polytomous analyses are also well suited to studies of chemical modes of action.

Among the reasons that this type of quantal response analysis has not been widely used in investigations with pesticides, the most important is probably the limited number of examples readily understandable to biological scientists. Another important deterrent has been the availability of few computer programs that can perform the required analyses. One such program was shown earlier. Using the multinomial program in R, we were able to analyze polytomous data without having to collapse the categorical effects. A user-friendly, well-documented polytomous probit or logit computer program is needed for routine use by biologists.

REFERENCES

1. Finney, D. J., *Probit Analysis*, Cambridge University Press, Cambridge, England, 1971.
2. Gurland, J., Lee, I., and Dahm, P. A., Polychotomous quantal response in biological assay, *Biometrics* 16, 382, 1960.
3. Hughes, G. A., Robertson, J. L., and Savin, N. E., Comparison of chemical effectiveness by the multinomial logit technique, *J. Econ. Entomol.* 80, 18, 1987.
4. Tukey, J. W., Ciminera, J. L., and Heyse, J. F., Testing the statistical certainty of a response to increasing doses of a drug, *Biometrics* 41, 295, 1985.
5. R version 1.9.1, Released June 21, 2004. See http://www.r-project.org.
6. Robertson, J. L. and Kimball, R. A., Effects of insect growth regulators on the western spruce budworm (*Choristoneura occidentalis*) (Lepidoptera: Tortricidae). I. Lethal effects of last instar treatments, *Can. Entomol.* 111, 1361, 1979.

CHAPTER 15

Improving Prediction Based on Dose–Response Bioassays

The problems that Paula has studied all concern estimation of response of laboratory populations. Precise estimation of response in a bioassay can be ensured by use of adequate sample sizes, careful attention to dose placement, and careful attention to sound experimental design. Any estimate from a laboratory bioassay analyzed by standard statistical models is experiment specific because it does not predict the results of a future experiment. Prediction statistics have been used in econometrics[1] and studies of travel demand[2] for many years; their use in ecotoxicology[3] emphasizes the effects of human activities on the environment. Available statistical methods for prediction[3] should be used, adapted, and, where necessary, developed for routine use in toxicological studies with invertebrates and other organisms besides humans.

15.1 ATTEMPTS TO IMPROVE METHODS

15.1.1 Exposure

Among the numerous methods of exposing test subjects to pesticides (e.g., see Busvine[4]), a technique similar to that used in a field application is usually chosen when a bioassay is intended to predict efficacy. For example, spray application devices (Potter[5] or Robertson et al.[6]) have been carefully designed and constructed to simulate the procedure used to apply pesticides to many phytophagous species.

Although use of any method that better simulates exposure makes an assay somewhat more realistic, the overall predictive value of the experiment does not necessarily increase. For example, if target insects are sprayed while fully exposed when they would be shielded by foliage in nature, results of the bioassay will obviously pertain only to fully exposed insects. Extrapolation of their responses to responses of insects in a natural situation is impossible. Thus, when Robertson and Boelter[7] applied sprays to fully exposed Douglas-fir tussock moth, the estimates of rates that they obtained bore no obvious relationship to those necessary for adequate efficacy against populations in the field.

An obvious modification of the direct spray method is spray application to insects inhabiting host plant foliage. This was the method that Dr. Maven used when she sprayed the tomatoes. This bioassay procedure increases the realism of exposure because the spatial arrangements of the target organisms resemble those in a natural environment. Some organisms may contact spray directly, but many will be at least partially shielded from the aerosol form of toxicant. Those protected may contact deposits on foliage or ingest the active ingredient after the spray has been applied. A series of such bioassays with western spruce budworm was used by Haverty and Robertson[8] to derive version factors that appear to be applicable for prediction of the LD_{90} of pesticide in the field based on the lethal dose in the laboratory bioassay.

At first reading, Dr. Maven thinks that this study provides the best guidance available about how to do a predictive bioassay. And it is, but only in part. What went wrong? First, the results concern only the last (sixth) instars of the insect. Use of only one developmental stage, selected from a laboratory colony to be of highly uniform age within the instar, negated the predictive value of the procedure with respect to a natural population.

15.1.2 Scoring Process

The scoring process most often used in pesticide bioassays is binary: dead vs. alive. More realistic scoring that includes effects between the extremes of life and death required the use of sophisticated statistical methods and have only recently been used for the analyses of large data sets from a bioassay (Hughes et al.[9]). These methods are described in Chapter 14.

In practice, however; investigations with this improved scoring process did not consider the effects on reproduction of survivors or survival in the F1 generation. Reproductive effects and survival of progeny have been documented for a few pesticides (e.g., Robertson and Kimball[10]), but such data have not been considered within the context of population dynamics.

Ahmadi[11] used life table parameters in conjunction with a conventional bioassay to demonstrate a procedure that he called demographic toxicology. However, concentrations of dicofol that he tested on twospotted spider mite, *Tetranychus urticae* Koch, were sublethal to adults and were tested only with residues on hatching eggs. Thus, effects on the parental generation were not included and the more refined scoring process failed to predict effects of the pesticide on a real population despite its inclusion of ecological parameters.

15.1.3 Significant Independent Variables

Laboratory bioassays provide an ideal means to test whether one or more specific variables significantly affect response to a pesticide. For example, temperature (e.g., Sparks et al.[12] and Schmidt and Robertson[13]), relative humidity and temperature (Reichenbach and Collins[14]), and body weight (Robertson et al.[15]) have been identified as significant factors in responses of several insect species to some pesticides.

The likelihood of resistance development (e.g., Cochran,[16] McGaughey and Johnson,[17] and Hoy and Ouyang[18]) can be estimated and suppositions about what might happen in nature can be made. However, just as modifications of the scoring process and use of more realistic prediction, identification of significant independent variables provides information that is difficult to relate to pesticide efficacy at the population level. From the examples given here and in earlier chapters (see Chapters 5 and 6), Dr. Maven knows that temperature, relative humidity, insect body weight, and genetic selection affect response of individuals, but the interactive roles of these variables in population response are still unknown.

15.1.4 Multiple Bioassays

Another approach to more realistic prediction has consisted of definition of separate aspects of pesticide effectiveness, followed by attempts to synthesize results into overall estimates of effectiveness on a population. Robertson and Haverty,[19] for example, defined effects by ingestion and contact, residual toxicities, and rainfastness in experiments with western spruce budworm. They devised a scoring system based on these four factors and predicted expected field effectiveness. Later field trials showed that the efficacy of two of the six pesticides had been grossly overestimated and the efficacy of another pesticide had been grossly underestimated (Markin and Johnson[20]). Penman et al. (1986) experienced similar difficulties with a multiple bioassay approach based on different behaviors of twospotted spider mite.

The underlying problem in the multiple bioassay approach is that the relative contributions of the various aspects of pesticide effectiveness to overall efficacy are unknown. For example, even Dr. Maven does not know whether contact toxicity is as important as toxicity by ingestion when chemical Y is applied to a population of *Patronius giganticus*. She simply has not a clue about how relative importance can be quantified.

15.1.5 Optimal Time of Application

One definition of optimal time of application is the time during population development when a pesticide will achieve maximum effect with minimum active ingredient. Early attempts to identify optimal time consisted of simple empirical estimates. First, the chemical was applied to each developmental stage (selected to be of uniform ages within the stage) and the toxicity of each stage was estimated. Based on relative toxicities (e.g., LD_{50} stage X $\div$ LD_{50}; most susceptible stage), an empirical estimate of optimal time was made. For example, if chemical X was most toxic to second instars, the optimal time of application would be when a population consisted mainly of second instars. Studies by Grannett and Retnakaran[21] and Robertson[22] demonstrated the use of such empirical estimates.

Robertson and Haverty[23] used a more refined procedure that was based on the instar distribution, over time, of a small number of western spruce worm. Dr. Maven notices some problems: Not only was the sample size small ($n = 100$), but also larvae were reared individually on artificial diet under highly controlled laboratory conditions. Next, Robertson et al.[24] developed a contour plot procedure to estimate optimal time of application. However, as discussed by Robertson et al.,[25] the estimation procedure lacked an adequate basis because optimal time was expressed in terms of chronological age of relatively few laboratory-reared larvae (i.e., the same 100 insects used by Robertson and Haverty[23] and cannot be equated with developmental rates of field populations, which may vary in instar composition as the result of factors as asynchronous hatch or different climatic conditions, mainly temperature).

15.1.6 Test Subjects

Another approach may be to collect and test samples of field populations in the laboratory. The age of individuals in the sample cannot be determined specifically, but the sample itself would represent individuals from a population of a particular age distribution relative to physiological time (e.g., the sample is of a population X degree-days from first egg hatch). The particular method used to describe a field population in terms of physiological time, whether degree-days (linear development rates) or proportion of development over time (nonlinear development rates), must be appropriate to the insect population sampled. Close cooperation with ecologists will therefore be required.

An alternative procedure would be to simulate selection of test subjects from a field population in the laboratory if population samples cannot be collected from the field because of economic or practical constraints (such as tree height). With this alternative, arthropods reared in the laboratory would constitute the population. Physiological time, rather than chronological time, would be used as the basis of selection of individuals for testing. Such a selection process should be equally applicable to routine activities, such as screening, as well as more complex problems, such as optimal time of pesticide application. The contour plot method[24] might be used to assess optimal time of application at minimum application rates, but with physiological time rather than chronological time as the means to describe the age of the population.

15.1.7 Reasons for Failure

A problem in predictive value may be the use of insects reared on artificial diets. Even if a bioassay predicted efficacy at the population level for a species fed artificial diet, would the predictive value of results be relevant to a population of the same species feeding on natural hosts? Evidence reviewed by Rose et al.[26] suggests that the answer might be no. For phytophagous insects, evidence that responses to pesticides are significantly affected by host plant species or their quality continues to accumulate. Resistance monitoring may even be confounded by the effects of different host plant species on response.[27,28]

At least for the generation being tested in the bioassay, we recommend that the population be fed natural substrates rather than artificial diet. For most phytophagous species, potted plants or a continuous supply of fresh host–plant material could be used. Artificial diet should be used in the bioassay only when the diet itself does not contribute significantly to response. The latter can be learned if several bioassays are conducted to compare response between individuals reared on artificial diet and those reared on host–plant material. In some cases, there are no significant differences.[29]

Once the arthropods have been exposed to the treatment that is as realistic a way as possible (e.g., spray application), monitoring of their development should continue throughout completion of the life cycle. An arbitrary evaluation period after treatment, such as seven days, should not be used because long-term effects are liable to be overlooked when in fact they are important at the population level. If possible, survivors should be mated and development of the F_1 generation monitored. Ideally, table parameters obtained from these data might be used to identify second-generation treatment effects that should otherwise go undetected. Because life table studies are so time and labor intensive, a simpler approach is to hold test subjects until they reach sexual maturity, allow a subsample to mate, and maintain the progeny until they too reach maturity. Fitness of treated survivors, untreated survivors, and their respective progeny can then be compared.

REFERENCES

1. Domencich, T. A. and McFadden, D., *Urban Travel Demand*, American Elsevier, New York, 1975.
2. Maddala, G. S., *Econometrics*, McGraw-Hill, New York, 1977.
3. Cairns, J. Jr. and Niederlehner, B. R., Predictive ecotoxicology, in Hoffman, D. J., Rattner, B. A., Burton, G. A. Jr., and Cairns, J. Jr., eds., *Handbook of Ecotoxicology*, CRC Press, Boca Raton, FL, 2003, pp. 911–924.
4. Busvine, J. R., *A Critical Review of Techniques for Testing Insecticides*, Commonwealth Agricultural Bureaux, London, 1971.
5. Potter, C., An improved laboratory apparatus for applying direct sprays and surface films, with data on the electrostatic charge on atomized spray fluids, *Ann. Appl. Biol.* 39, 1, 1952.
6. Robertson, J. L., Lyon, R. L., Andrews, T. L., Moellman, E. E., and Page, M., Moellman spray chamber: Versatile research tool for laboratory bioassays, USDA Forest Service Res., Note, PSW-335, 1979.
7. Robertson, J. L. and Boelter, L. M., Toxicity of insecticides to Douglas-fir tussock moth *Orgyia pseudotsugata* (Lepidoptera: Lymantriidae). I. Contact and feeding toxicity, *Can. Entomol.* 111, 114, 1979.
8. Haverty, M. I. and Robertson, J. L., Laboratory bioassays for selecting candidate insecticides and application rates for field tests on the western spruce budworm, *J. Econ. Entomol.* 75, 179, 1982.
9. Hughes, G. A., Robertson, J. L., and Savin, N. E., Comparison of chemical effectiveness by the multinomial logit technique, *J. Econ. Entomol.* 80, 18, 1987.
10. Robertson, J. L. and Kimball, R. A., Effect of insect growth regulators on the western spruce budworm (*Choristoneura occidentalis*) (Lepidoptera: Tortricidae). II. Fecundity and fertility reduction following last instar treatments, *Can. Entomol.* 111, 1369, 1979.
11. Ahmadi, A., Demographic toxicology as a method for studying the dicofol-twospotted spider mite (Acari: Tetranychidae) system, *J. Econ. Entomol.* 76, 239, 1983.
12. Sparks, T. C., Shour, M. H., and Wellenyer, E. D., Temperature-toxicity relationships of pyrethroids on three lepidopterans, *J. Econ. Entomol.* 75, 643, 1982.

13. Schmidt, C. D. and Robertson, J. L., Effects of treatment technique on response of horn flies (Diptera: Muscidae) to permethrin at different temperatures, *J. Econ. Entomol.* 79, 684, 1986.

14. Reichenbach, N. E. and Collins, W. J., Multiple logit analyses of the effects of temperature and humidity on the toxicity of propoxur to German cockroaches (Orthroptera: Blattellidae) and western spruce budworm larvae (Lepidoptera; Tortricidae), *J. Econ. Entomol.* 77, 31, 1984.

15. Robertson, J. L., Savin, N. E., and Russell, R. M., Weight as a variable in response of western spruce budworm to insecticides, *J. Econ. Entomol.* 74, 643, 1981.

16. Cochran, D. G., Selecton for pyrethroid resistance in the German cockroach (Dictyoptera: Blatellidae), *J. Econ. Entomol.* 80, 1122, 1987.

17. McGaughey, W. H. and Johnson, D. W., Toxicity of different serotypes of toxins of *Bacillus thuringiensis* to resistant and susceptible Indianmeal moth (Lepidoptera: Pyralidae), *J. Econ. Entomol.* 80, 1122, 1987.

18. Hoy, M. A. and Ouyang, Y. L., Selection of the western predatory mite, *Metaseiulus occidentalis* (Acari: Pytoseiidae), for resistance to abamectin, *J. Econ. Entomol.* 82, 35, 1989.

19. Robertson, J. L. and Haverty, M. I., Multiphase laboratory bioassays to select chemicals for field testing on western spruce budworm, *J. Econ. Entomol.* 74, 148, 1981.

20. Markin, G. P. and Johnson, D. R., Western spruce budworm aerial field test, *Insecticide Acaricide Tests* 7, 203, 1982.

21. Grannett, J. and Retnakaran, A., Stradial susceptibility of eastern spruce budworm (*Choristoneura fumiferana*) (Lepidoptera: Tortricidae) to the insect growth regular Dimilin, *Can. Entomol.* 109, 893, 1977.

22. Robertson, J. L., Contact and feeding toxicities of acepahte and carbaryl to larval stages of the western spruce budworm, *Choristoneura occidentalis* (Lepidoptera: Tortricidae), *Can. Entomol.* 112, 1001, 1980.

23. Robertson, J. L. and Haverty, M. L., Estimation of rates and times of application for selected insect growth regular formulations to western spruce budworm, *J. Ga. Entomol. Soc.* 17, 297, 1982.

24. Robertson, J. L., Richmond, C. E., and Preisler, H. K., Lethal and sublethal effects of avermectin B1 on the western spruce budworm (Lepidoptera: Tortricidae), *J. Econ. Entomol.* 78, 1129, 1985.

25 Robertson, J. L., Worner, S. P., and Preisler, H. K., Comparative optimal times of application of benzoylphenylureas to western spruce budworm, *Choristoneura occidentalis* (Lepidoptera: Tortricidae), *Can. Entomol.* 121, 75, 1989.

26. Rose, R. L., Sparks, T. C., and Smith, C. M., Insecticide toxicity to the soybean looper and velvelbean caterpillar larvae (Lepidoptera: Noctuidae) as influenced by feeding on resistant soybean (PI 227687) and coumestrol, *J. Econ. Entomol.* 81, 1288, 1988.

27. Robertson, J. L., Armstrong, K. F., Suckling, D. M., and Preisler, H. K., Effects of host plants on the toxicity of azinphosmethyl to susceptible and resistant light brown apple moth (Lepidoptera: Tortricidae), *J. Econ. Entomol.* 83, 2124, 1990.

28. Seigfried, B. D. and Mullin, C. A., Influence of alternative host plant feeding on Aldrin susceptibility and detoxification enzymes in western and northern corn rootworms, *Pestic. Biochem. Physiol.* 35, 155, 1989.

29. Jones, M. M., Robertson, J. L., and Weinzierl, R. A., Susceptibility of Oriental fruit moth (Lepidoptera: Tortricidae) larvae to selected reduced-risk insecticides, *J. Econ. Entomol.* 103, 5, 1815, 2010.

CHAPTER 16

Population Toxicology

The major deficiency in laboratory bioassays that has destroyed their predictive value has been failure to consider effects of pesticides at the level of an arthropod population. Instead, investigators have expended a great deal of effort to select highly uniform individuals as test subjects. Regardless of number, responses of these individuals cannot be added, averaged, or otherwise manipulated to predict a population's response. Why? Because their ability does not reflect that of a population.

The role of variability in rates of population development has been ignored in laboratory bioassays done to predict efficacy. A population in nature does not, for example, consist of individuals whose developmental rates and physical characteristics are highly uniform. A new approach is clearly needed for laboratory bioassays to approximate responses likely to occur in a field population. Robertson and Worner[1] suggest that population response, rather than response of individuals selected for their uniform characteristics, must be emphasized.

The very precision that ensures the validity of bioassays done to test the significance of selected variables, relationships between chemical structure and activity, and other special purposes negates the predictive value of the experiment with regard to a population. Even when multiple bioassay techniques are used, prediction is impossible because we do not know how various facets of effectiveness should be put together to define overall efficacy. Finally, use of a more ecological approach involving both pesticide efficacy and life table parameters misses the point when only one life stage is treated and sublethal application rates are used.

Researchers have probably not considered a population approach to laboratory bioassays because the definition of population age is difficult and analyses of results for anything besides instar categories of uniform age and weight present another problem. Numerous methods to define the physiological age distribution of a population have been described. These range from simple degree day approaches to complex distributional models.[2,3] While the accuracy of these methods may be questionable, they at least provide a means to describe the physiological age of a population relative to some measurable, biologically significant, point in time (e.g., first egg deposition, termination of diapause). This method might also be a means to begin studying the problem of prediction of response once an arthropod population is treated with a pesticide. Similarly, the effects of treatment on a developing population can now be assessed using ridit analysis.[4]

Samples of field populations of the target arthropod species might be collected and tested in the laboratory. One may choose to sample all adults or immatures regardless of age, or there are methods to reduce the age continuum (e.g., the sample is of a population X degree-days from first egg hatch). When determining the sampling method to use regarding physiological time, it must be appropriate for the field population being sampled whether to use degree-days or proportion of development over time.

Precedents for evaluation of efficacy in relation to insect phenology are found in the literature concerning tests with chemicals in the field. Walker et al.[5] described the efficacy of pesticide spray applications for California red scale, *Aonidiella aurantii* (Maskell), relative to male flight phenology. Similar studies[6,7] have been reported with San Jose scale, *Quadraspidiotis perniciosus* (Comstock). Just as these investigations, all of which were based on physiological development of populations, provided reliable estimates of timing of the application of chemicals for control, predictions of timing of pesticide application for maximum efficacy should be possible if laboratory bioassays are based on population age.

If population samples cannot be collected from the field because of time or practical constraints, a laboratory population could be used to simulate selection of test subjects from a field population. Physiological time would be used as a means to describe the age of the population. This information would then be used to screen individuals or determine the optimal timing of pesticide applications.

Robertson and Worner[1] suggested, at least for the generation being tested, that the population be fed natural substrates rather than artificial diet. For most phytophagous species, potted plants or a continuous supply of fresh host plant material can be used. Dr. Maven agrees: artificial diet should not be used unless response to pesticides has been shown not to be affected by the diet.

Once the arthropod species has been exposed to the pesticide in as realistic way as possible (e.g., spray application), monitoring of their development should continue throughout completion of their life cycle. An arbitrary evaluation period after treatment, such as 7 days, should not be used because long-term effects are subject to being overlooked, when in fact they may be important at the population level. If possible, survivors should be mated and development of the F_1 generation monitored.

Ideally, life table parameters obtained from these data might be used to identify second-generation treatment effects that would otherwise go undetected. Because life table studies are so time and labor intensive, a simpler approach is to hold test subjects until they reach sexual maturity, allow a sample to mate, and maintain the progeny until they too reach maturity. Fitness of treated survivors, untreated survivors, and their respective progeny might then be compared.

Kogan[9] has emphasized the need to base pest management decisions in the field on sound principles of population ecology. Robertson and Worner[1] and Stark and Banks[10] have emphasized the same concerns for laboratory bioassays done to predict pesticide efficacy to populations in the field. Costs of doing population toxicology may exceed the expense presently incurred in doing a laboratory bioassay, but the reliability of the resultant predictions would be well worth both the extra time and money involved.

REFERENCES

1. Robertson, J. L. and Worner, S. P., Population toxicology: Suggestions for laboratory bioassays to predict pesticide efficacy, *J. Econ. Entomol.* 83, 1, 1990.
2. Hudes, E. S. and Shoemaker, C. A., Inferential method for modelling insect phenology and its application to the spruce budworm (Lepidoptera: Tortricidae), *Environ. Entomol.* 17, 97, 1988.
3. Wagner, T. L., Wu, H., Sharpe, P. J. H., and Coulson, R. N., Modelling distributions of insect development time: A literature review and application of the Weibull function, *Ann. Entomol. Soc. Am.* 77, 475, 1984.
4. Howard, P. J. A. and Howard, D. M., The application of ridit analysis to phenological observations, *J. Appl. Stat.* 12, 29, 1985.
5. Walker, G. P., Aitken, D. C., O'Connell, N. V., and Smith, D., Using phenology to time insecticide applications for control of California red scale (Homoptera: Diaspididae) on citrus, *J. Econ. Entomol.* 83, 189, 1990.
6. Rice, R. E. and Jones, R. A., Timing post-bloom sprays for peach tree twig borer (Lepidoptera: Gelichiidae) and San Jose scale (Homoptera: Diaspididae), *J. Econ. Entomol.* 81, 293, 1988.
7. Downing, R. S. and Logan, D. M., A new approach to San Jose scale control (Hemiptera: Diaspidae), *Can. Entomol.* 109, 1249, 1977.
8. Robertson, J. L., Richmond, C. E., and Preisler, H. K., Lethal and sublethal effects of avermectin B_1 on the western spruce budworm (Lepidoptera: Tortricidae), *J. Econ. Entomol.* 78, 1129, 1985.
9. Kogan, M., ed., *Ecological Theory and Integrated Pest Management Practice*, Wiley Interscience, New York, 1986.
10. Stark, J. D. and Banks, J. E., Population-level effects of pesticides and other toxicants on arthropods, *Annu. Rev. Entomol.* 48, 505, 2003.

Index